Name _______________________ Class _____________ Date _____________

Directed Reading

Section: Passive Transport

Read each question, and write your answer in the space provided.

1. What is passive transport? Why is diffusion an example of passive transport?

____

____

____

____

2. How does the cell membrane help cells maintain homeostasis?

____

____

____

3. What determines the direction in which a substance diffuses across a membrane?

____

____

____

____

4. Describe the state of equilibrium.

____

____

In the space provided, explain how the terms in each pair differ in meaning.

5. osmosis, diffusion

____

____

____

6. hypertonic solution, hypotonic solution

____

____

____

7. isotonic solution, equilibrium

____

____

____

____

In the space provided, write the letter of the description that best matches the term or phrase.

_______ **8.** hypertonic solution

_______ **9.** selective permeability

_______ **10.** osmosis

_______ **11.** negatively charged

_______ **12.** facilitated diffusion

_______ **13.** concentration gradient

_______ **14.** ion channel

a. difference in the concentration of a substance across a space

b. the inside of a typical cell

c. diffusion of water through a cell membrane

d. allows charged molecules to pass through the cell membrane

e. enables a cell to control what enters and leaves

f. will cause a cell to shrivel up

g. involves carrier proteins

Directed Reading

Section: Active Transport

Complete each statement by writing the correct term or phrase in the space provided.

1. The transport of a substance across the cell membrane against its concentration gradient is called ______________________ ______________________ .

2. Active transport requires the cell to use ______________________ .

3. The energy needed for active transport is usually supplied by

______________________ .

4. The sodium-potassium pump is a(n) ______________________ protein.

5. The concentration of sodium ions inside the cell is usually

______________________ than the concentration of sodium ions

outside the cell.

6. The concentration of potassium ions inside the cell is usually

______________________ than the concentration of potassium

ions outside the cell.

7. The sodium-potassium pump picks up ______________________ ions outside the cell.

8. The sodium-potassium pump releases ______________________ ions inside the cell.

Read each question, and write your answer in the space provided.

9. Explain why proteins and polysaccharides cannot diffuse through the membrane like water does.

____

____

____

Directed Reading *continued*

10. What is the difference between endocytosis and exocytosis?

____

____

11. How is a vesicle formed in endocytosis?

____

____

12. What happens to a vesicle in exocytosis?

____

____

13. How do sodium-potassium pumps support the efficient functioning of cells?

____

____

____

In the space provided, write the letter of the description that best matches the term or phrase.

_______ **14.** signal molecule

_______ **15.** receptor protein

_______ **16.** ion channel

_______ **17.** second messenger

_______ **18.** enzyme action

_______ **19.** beta blocker

_______ **20.** changes in permeability

a. a large protein in the cell membrane that transports a specific ion

b. acts as a signal molecule in the cytoplasm

c. a protein that binds to a specific signal molecule

d. speeds up chemical reactions in the cell

e. a drug that interferes with the binding of signal molecules to receptor proteins in heart muscles

f. carries information throughout the body and to other cells

g. occur when a receptor protein is coupled with an ion channel

Skills Worksheet

Active Reading

Section: Passive Transport

Read the passage below. Notice that the sentences are numbered. Then answer the questions that follow.

[1] The diffusion of water through a selectively permeable membrane is called **osmosis**. [2] Like other forms of diffusion, osmosis involves the movement of a substance—water—down its concentration gradient. [3] Osmosis is a type of passive transport.

[4] If the solutions on either side of the cell membrane have different concentrations of dissolved particles, they will also have different concentrations of "free" water molecules. [5] Osmosis will occur as water molecules diffuse into the solution with the lower concentration of free water molecules.

SKILL: READING EFFECTIVELY

Read each question, and write your answer in the space provided.

1. What key term is defined in this passage? What does this term mean?

____

____

2. How are diffusion and osmosis related?

____

____

3. What does the word *water* in Sentence 2 tell you about osmosis?

____

____

In the space provided, write the letter of the term or phrase that best completes the statement.

_______ **4.** Osmosis is a type of
    a. passive transport.
    b. diffusion.
    c. active transport.
    d. Both (a) and (b)

Active Reading

Section: Active Transport

Read the passage below. Then answer the questions that follow.

The movement of a substance into a cell by a vesicle is called **endocytosis**. During endocytosis, the cell membrane forms a pouch around a substance outside the cell. The pouch then closes up and pinches off from the membrane to form a vesicle. Vesicles formed by endocytosis may fuse with lysosomes or other organelles.

The movement of a substance by a vesicle to the outside of a cell is called **exocytosis**. During exocytosis, vesicles in the cell fuse with the cell membrane, releasing their contents. Cells use exocytosis to export proteins that are modified by the Golgi apparatus. Nerve cells and cells of various glands, for example, release proteins by exocytosis.

SKILL: RECOGNIZING SIMILARITIES AND DIFFERENCES

Complete the table below. In the first column, write two characteristics of cells in endocytosis. In the second column, write two characteristics of cells in exocytosis.

Endocytosis	Exocytosis
1.	3.
2.	4.

Active Reading *continued*

Read the question, and write your answer in the space provided.

5. The prefix *endo-* means "inside or within." How would knowing this prefix meaning help you define the key term *endocytosis*?

____

____

____

____

In the space provided, write the letter of the term or phrase that best completes the statement.

______ **6.** Through the process of exocytosis, nerve cells
   a. form pouches.
   b. release proteins.
   c. fuse with lysosomes.
   d. Both (a) and (b)

Skills Worksheet

Vocabulary Review

In the space provided, write the letter of the description that best matches the term or phrase.

_______ 1. passive transport

_______ 2. concentration gradient

_______ 3. equilibrium

_______ 4. diffusion

_______ 5. osmosis

_______ 6. hypertonic solution

_______ 7. hypotonic solution

_______ 8. isotonic solution

_______ 9. ion channel

_______ 10. carrier protein

_______ 11. facilitated diffusion

_______ 12. active transport

_______ 13. sodium-potassium pump

_______ 14. endocytosis

_______ 15. exocytosis

_______ 16. receptor protein

_______ 17. second messenger

a. movement of a substance down the substance's concentration gradient

b. causes a cell to shrink because of osmosis

c. movement of a substance by a vesicle to the outside of a cell

d. carrier protein used in active transport

e. protein used to transport specific substances

f. transport protein through which ions can pass

g. movement of a substance by a vesicle to the inside of a cell

h. does not require energy from the cell

i. concentration of a substance is equal throughout a space

j. difference in the concentration of a substance across a space

k. diffusion of water through a selectively permeable membrane

l. causes a cell to swell because of osmosis

m. passive transport using carrier proteins

n. produces no change in cell volume because of osmosis

o. movement of a substance against the substance's concentration gradient

p. acts as a signal molecule in the cytoplasm

q. binds to a signal molecule, enabling the cell to respond to the signal molecule

Science Skills

Predicting

Use the information below and the figure at right to answer questions 1–3.

EXPERIMENT A

A selectively permeable membrane separates the solutions in the arms of the U-tube shown at right. The membrane is permeable to water and to substance A but not to substance B. Forty grams of substance A and 20 g of substance B have been added to the water on side 1 of the U-tube. Twenty grams of substance A and 40 g of substance B have been added to the water on side 2 of the U-tube. Assume that after a period of time, the solutions on either side of the membrane have reached equilibrium.

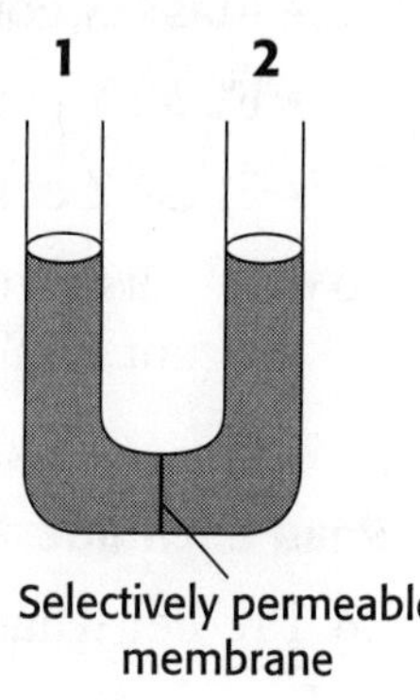

Read each question, and write your answer in the space provided.

1. How many grams of substance A will be in solution on side 1 of the U-tube? How many grams of substance A will be in solution on side 2? Explain.

____

____

____

____

____

2. How many grams of substance B will be in solution on side 1 of the U-tube? How many grams of substance B will be in solution on side 2? Explain.

____

____

____

3. What has happened to the water level in the U-tube? Explain.

____

____

____

____

Use the information below to answer questions 4–6.

EXPERIMENT B

The cell membrane of red blood cells is permeable to water but not to sodium chloride, NaCl. Suppose that you have three flasks:

- Flask X contains a solution that is 0.5 percent NaCl.
- Flask Y contains a solution that is 0.9 percent NaCl.
- Flask Z contains a solution that is 1.5 percent NaCl.

To each flask, you add red blood cells, which contain a solution that is 0.9 percent NaCl.

Read each question, and write your answer in the space provided.

4. Predict what will happen to the red blood cells in flask X.

____

____

____

____

5. Predict what will happen to the red blood cells in flask Y.

____

____

____

____

6. Predict what will happen to the red blood cells in flask Z.

____

____

____

____

Concept Mapping

Using the terms and phrases provided below, complete the concept map showing the characteristics of cell transport.

active transport

concentration gradients

endocytosis

facilitated diffusion

ion channels

passive transport

receptor proteins

sodium-potassium pump

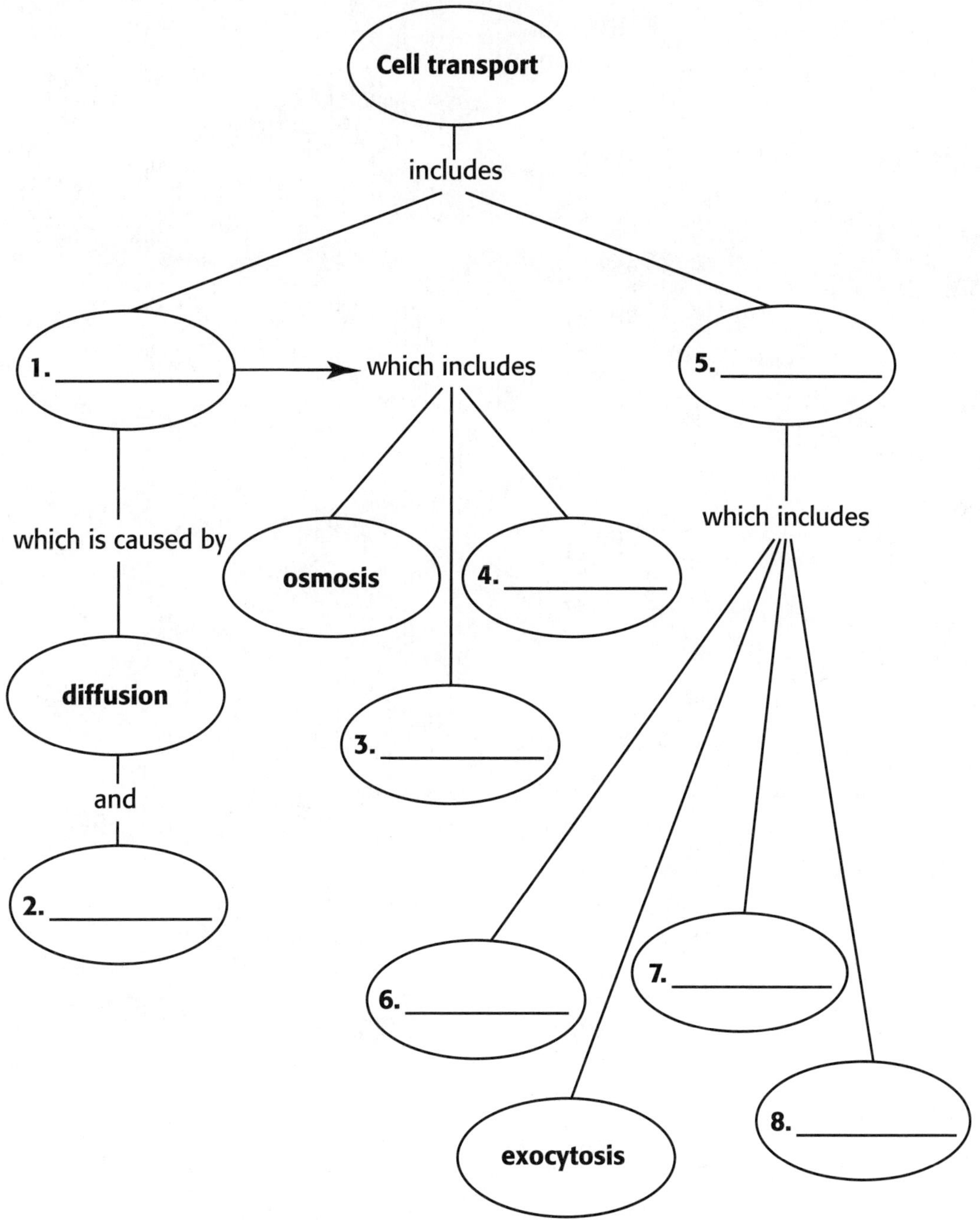

Critical Thinking

Look-Alikes

In the space provided, write the letter of the term or phrase that best describes how each numbered item looks.

_______ **1.** random-motion-of-molecule model

_______ **2.** diffusion

_______ **3.** ion channel

a. a crowd of people filling a room

b. a long tube

c. millions of table tennis balls bouncing into one another

Work-Alikes

In the space provided, write the letter of the term or phrase that best describes how each numbered item functions.

_______ **4.** carrier protein

_______ **5.** sodium-potassium pump

_______ **6.** endocytosis

_______ **7.** exocytosis

_______ **8.** receptor protein

a. language interpreter

b. immigrants

c. train leaving the station

d. football player carrying the ball

e. bailing water out of a sinking boat

Trade-offs

In the space provided, write the letter of the bad news item that best matches each numbered good news item below.

Good News	**Bad News**

_______ **9.** Very small, nonpolar molecules can diffuse across a cell membrane.

_______ **10.** Diffusing down a concentration gradient is easy.

_______ **11.** Receptor proteins receive messages carried by signal molecules.

_______ **12.** The sodium-potassium pump moves sodium out of the cell.

a. Most polar molecules are repelled from crossing the membrane.

b. It diffuses back in.

c. Illegal drugs can block them.

d. Diffusing up a concentration gradient is difficult.

Linkages

In the spaces provided, write the letters of the two terms or phrases that are linked together by the term or phrase in the middle. The choices can be placed in any order.

13. _______ diffusion _______

14. _______ osmosis _______

15. _______ sodium potassium pump _______

16. _______ isotonic _______

17. _______ carrier protein _______

18. _______ receptor proteins _______

a. hypotonic

b. high concentrations of ions

c. high concentration of water

d. signal molecule

e. hypertonic

f. low concentration of water

g. ions released

h. phosphate group binds to pump

i. second messenger

j. low concentration of ions

k. changes shape

l. facilitated diffusion

Analogies

An analogy is a relationship between two pairs of terms or phrases written as a : b :: c : d. The symbol : is read as "is to," and the symbol :: is read as "as." In the space provided, write the letter of the pair of terms or phrases that best completes the analogy shown.

_______**19.** active transport : ATP ::
  a. carbon : fuel
  b. passive transport : sodium
  c. facilitated diffusion : concentration
  d. campfire : wood

_______**20.** exocytosis : vesicles out ::
  a. endocytosis : exocytosis
  b. endocytosis : vesicles in
  c. exocytosis : ATP
  d. vesicles : ATP

_______**21.** people : from newspapers ::
  a. second messenger : from enzymes
  b. signal molecules : from receptor proteins
  c. signal molecules : from enzymes
  d. receptor proteins : from signal molecules

Skills Worksheet

Test Prep Pretest

In the space provided, write the letter of the term or phrase that best completes each statement or best answers each question.

_______ **1.** When a receptor protein in a cell membrane acts as an enzyme, the receptor protein
    a. changes its shape to allow the signal molecule to enter the cell.
    b. causes chemical changes in the cell.
    c. activates a second messenger that acts as a signal molecule within the cell.
    d. changes the permeability of the cell membrane.

_______ **2.** Which of the following is NOT a characteristic of an ion channel?
    a. It extends from one side of the cell membrane to the other.
    b. It may or may not have a gate.
    c. It is polar, so charged substances, such as ions, can pass through the nonpolar lipid bilayer.
    d. It allows ions to move against their concentration gradient.

_______ **3.** When a cell uses energy to transport a particle through the cell membrane to an area of higher concentration, the cell is using
    a. diffusion.   **c.** osmosis.
    b. active transport.   **d.** facilitated diffusion.

_______ **4.** The excretion of materials to the outside of a cell by discharging them from vesicles is called
    a. exocytosis.   **c.** osmosis.
    b. endocytosis.   **d.** diffusion.

_______ **5.** The mechanism that prevents sodium ions from building up inside the cell is called
    a. the sodium-potassium pump.   **c.** diffusion.
    b. endocytosis.   **d.** exocytosis.

Complete each statement by writing the correct term or phrase in the space provided.

Question 6 refers to the figure at right.

6. The process shown in the figure

   is _______________________ .

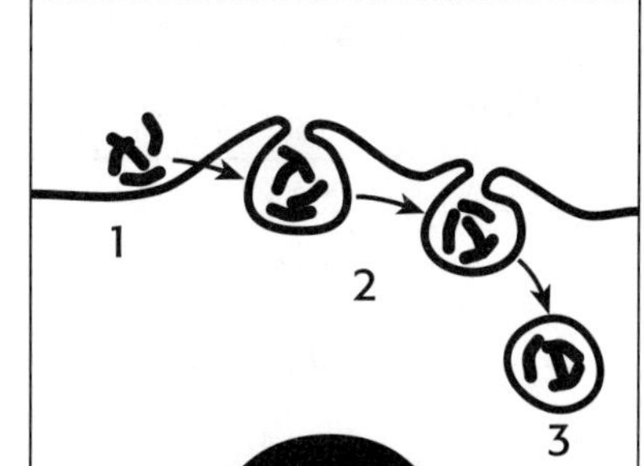

| Test Prep Pretest *continued*

---

7. Cell-surface proteins allow a cell to _________________ with other cells.

8. The _________________ _________________

_________________ requires energy to function.

9. When a substance moves from an area of low concentration to an area of

higher concentration, the substance moves _________________ its concentration gradient.

10. The movement of particles down their concentration gradient through carrier

proteins is known as _________________ _________________ .

11. A(n) _________________ _________________ amplifies the communication from a signal molecule.

12. A(n) _________________ _________________

_________________ in the cell membrane may be opened or closed.

Questions 13–15 refer to the figures below.

A   B   C

13. Figure *A* illustrates a cell in a(n) _________________ solution.

14. Figure *B* illustrates a cell in a(n) _________________ solution.

15. Figure *C* illustrates a cell in a(n) _________________ solution.

Read each question, and write your answer in the space provided.

16. Describe the electrical charge inside and outside a typical cell. Then explain how this affects an ion's ability to move into the cell.

____

____

____

____

Test Prep Pretest *continued*

17. Suppose you want to explain a concentration gradient to someone. Create a scenario that illustrates passive transport down the concentration gradient.

____

____

____

____

18. Using your understanding of osmosis, describe why putting salt on a pork chop before cooking it on a grill is likely to result in a dry, tough piece of meat.

____

____

____

____

19. How is facilitated diffusion different from the other passive transport processes?

____

____

____

20. How does a cell consume a food particle that is too large to pass through a protein channel?

____

____

____

Quiz

Section: Passive Transport

In the space provided, write the letter of the description that best matches the term or phrase.

_______ **1.** osmosis

_______ **2.** carrier protein

_______ **3.** equilibrium

_______ **4.** contractile vacuole

a. forces excess water out of a cell

b. an equal number of balls between two rooms

c. movement of a water down its concentration gradient

d. transports sugars and amino acids down their concentration gradient

In the space provided, write the letter of the term or phrase that best completes each statement or best answers each question.

_______ **5.** If the "free" water molecule concentration outside a cell is higher than that inside the cell, the solution outside of the cell is
    a. isotonic.
    b. hypertonic.
    c. hypotonic.
    d. None of the above

_______ **6.** If someone spills perfume in one room, people can soon smell it in a nearby room. Which of the following is this an example of?
    a. facilitated diffusion
    b. osmosis
    c. diffusion
    d. active transport

_______ **7.** Molecules that can diffuse across the membrane include
    a. many polar molecules.
    b. many nonpolar molecules.
    c. sugars
    d. amino acids.

_______ **8.** Ions function in your body to help
    a. cells generate body heat.
    b. the stomach digest food.
    c. cells send hormonal signals.
    d. nerve cells send electrical signals.

_______ **9.** The diffusion of ions across the membrane is influenced by which of the following?
    a. the electrical charge of the ion
    b. only the concentration gradient of the ion
    c. the number of enzymes in the cell membrane
    d. Both (a) and (c)

_______ **10.** Ion channels may open or close in response to
    a. stretching of the membrane.
    b. a change in electrical charge.
    c. the binding of a molecule to the ion channel.
    d. All of the above

Name _________________________________ Class _______________ Date _____________

Quiz

Section: Active Transport

In the space provided, write the letter of the description that best matches the term or phrase.

_______ **1.** signal molecule

_______ **2.** endocytosis

_______ **3.** beta-blocker

_______ **4.** active transport

_______ **5.** ATP

a. cell membrane forms a pouch around a substance

b. supplies the energy for active transport

c. carries information between cells

d. interferes with the binding of signal molecules

e. transport of a substance against its concentration gradient

In the space provided, write the letter of the term or phrase that best completes each statement or best answers each question.

_______ **6.** Which of the following is an example of active transport?
    a. equilibrium   **c.** facilitated diffusion
    b. sodium-potassium pump   **d.** osmosis

_______ **7.** The sodium-potassium pump transports
    a. sodium ions out of the cell.
    b. sodium ions into the cell.
    c. potassium ions out of the cell.
    d. Both (b) and (c)

_______ **8.** Without sodium-potassium pumps, cells
    a. might burst.
    b. might not be able to transport sugars as efficiently.
    c. would shrivel up.
    d. Both (a) and (b)

_______ **9.** Which of the following substances are too large for carrier proteins?
    a. ions
    b. glucose
    c. polysaccharides
    d. None of the above

_______ **10.** When a signal molecule binds to a receptor,
    a. it activates the sodium-potassium pump.
    b. the permeability of the receiving cell membrane may be changed.
    c. the signal molecule catalyzes an enzymatic reaction.
    d. All of the above

Chapter Test

Cells and Their Environment

In the space provided, write the letter of the description that best matches the term or phrase.

_______ **1.** sodium-potassium pump

_______ **2.** hypertonic solution

_______ **3.** isotonic solution

_______ **4.** facilitated diffusion

_______ **5.** exocytosis

_______ **6.** phospholipids

_______ **7.** gated ion channel

_______ **8.** hypotonic solutions

a. helps a cell rid itself of wastes

b. double layer that makes up cell membrane

c. may be open or closed

d. causes a cell to shrivel

e. transports potassium ions into the cell

f. transport of a specific substance down its concentration gradient by a carrier protein

g. may cause a cell to burst

h. should have no affect on a cell

In the space provided, write the letter of the term or phrase that best completes each statement or best answers each question.

_______ **9.** In the cell membrane, ion channels serve as
    a. food molecules.
    b. cell identifiers.
    c. information receivers.
    d. passageways.

_______ **10.** The diffusion of water through a selectively permeable membrane is called
    a. exocytosis.
    b. osmosis.
    c. active transport.
    d. endocytosis.

_______ **11.** Which of the following is NOT a characteristic of active transport?
    a. It moves substances against a concentration gradient.
    b. It requires energy from the cell.
    c. It involves facilitated diffusion.
    d. It relies on carrier proteins that often function as pumps.

_______ **12.** Diffusion is the movement of a substance
    a. through only a lipid bilayer.
    b. from an area of low concentration to an area of higher concentration.
    c. only in liquids.
    d. from an area of high concentration to an area of lower concentration.

______**13.** Which type of membrane protein transmits information into the cell by responding to signal molecules?
    a. carrier protein
    c. marker protein
    b. receptor protein
    d. None of the above

______**14.** An ion channel is a transport protein that
    a. moves substances against a concentration gradient.
    b. pumps ions only out of a cell.
    c. allows ions to move across the cell membrane so that the ions do not come in contact with the nonpolar interior of the lipid bilayer.
    d. has pores that are always open.

______**15.** When particles move out of a cell through facilitated diffusion, the cell
    a. gains energy.
    c. first gains and then uses energy.
    b. uses energy.
    d. does not use energy.

______**16.** The sodium-potassium pump
    a. helps cells maintain homeostasis.
    b. carries ions across the membrane from an area of high concentration to an area of lower concentration.
    c. helps maintain a sodium ion gradient, which cells use to transport sugars across the membrane.
    d. Both (a) and (c)

______**17.** Molecules that are too large to be moved through the cell membrane can be transported into the cell by
    a. osmosis.
    c. exocytosis.
    b. endocytosis.
    d. diffusion.

______**18.** Which of the following occurs when a signal molecule binds to a receptor protein on a cell's surface?
    a. The receptor can open an ion channel in the cell membrane.
    b. The receptor can act as an enzyme, causing chemical changes in the cytoplasm.
    c. The receptor can cause the formation of a second messenger.
    d. All of the above

______**19.** If the concentration of a sugar solution is lower outside the cell than inside the cell, which of the following will happen by osmosis?
    a. Sugar will move into the cell.
    b. Water will move into the cell.
    c. Sugar will move out of the cell.
    d. Water will move out of the cell.

______**20.** During which of the following processes do vesicles sometimes fuse with lysosomes?
    a. facilitated diffusion
    c. endocytosis
    b. osmosis
    d. exocytosis

Chapter Test

Cells and Their Environment

In the space provided, write the letter of the term or phrase that best completes each statement or best answers each question.

______ **1.** Which of the following is an example of osmosis?
   a. the movement of ions from an area of high concentration to an area of lower concentration
   b. the movement of ions from an area of low concentration to an area of higher concentration
   c. the movement of "free" water molecules from an area of high concentration to an area of lower concentration
   d. the movement of "free" water molecules from an area of low concentration to an area of higher concentration

Questions 2–5 refer to the figure below, which shows transport through the cell membrane.

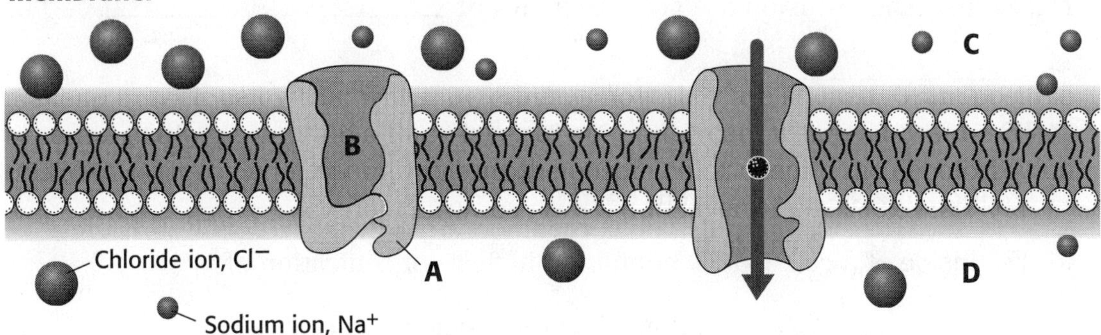

______ **2.** The structure labeled *A* is
   a. a gated ion channel.
   b. a sodium-potassium pump.
   c. an active transport protein.
   d. a messenger protein.

______ **3.** The interior of the pore, labeled *B*, enables substances to pass through the cell membrane without coming in contact with
   a. the solution outside the cell.
   b. the nonpolar interior of the lipid bilayer.
   c. the polar interior of the lipid bilayer.
   d. All of the above

______ **4.** The area that has the greatest concentration of ions in solution is labeled
   a. *A.*
   b. *B.*
   c. *C.*
   d. *D.*

______ **5.** What type of transport is illustrated by the figure above?
   a. osmosis
   b. active transport
   c. facilitated diffusion
   d. diffusion

In the space provided, write the letter of the description that best matches the term or phrase.

_______ **6.** equilibrium

_______ **7.** homeostasis

_______ **8.** second messenger

_______ **9.** exocytosis

_______ **10.** carrier protein

a. shields molecule from interior of lipid bilayer

b. used to export proteins

c. organisms adjust internally to changing external conditions

d. concentration of a substance is equal throughout a space

e. amplifies a signal inside a cell

Complete each statement by writing the correct term or phrase in the space provided.

11. The processes of endocytosis and exocytosis both require

_______________________ .

12. Osmosis and diffusion are both examples of _____________________

_______________________.

13. When the concentration of dissolved particles is the same throughout a

solution, the system is said to be in _________________________ .

14. The inside of a cell usually contains a higher concentration of

_________________________ ions than the outside of a cell.

15. A cell membrane is _________________________ permeable because it allows the passage of some substances and not others.

16. Using energy to transport molecules through a membrane from an area of low concentration to an area of higher concentration is called

_________________________ _________________________ .

17. In facilitated diffusion, _________________________ _________________________ move substances down their concentration gradient.

18. Substances outside a cell are transported into the cell by vesicles during

_________________________ .

19. A specific signal molecule binds to a(n) _________________________

_________________________ , causing changes in the cell.

20. A cell shrinks when it is placed in a(n) _________________________ solution.

| Chapter Test *continued*

Read each question, and write your answer in the space provided.

21. Why is it dangerous for humans to drink sea water?

____

____

____

____

22. Explain how the sodium-potassium pump works and why it is important.

____

____

____

____

____

____

____

23. Describe three ways in which the binding of a signal molecule to a receptor protein can change the activities of the receiving cell.

____

____

____

____

____

____

▌Chapter Test *continued*

24. Distinguish between diffusion and active transport.

____

____

____

____

____

25. Describe how different kinds of cells react to hypertonic solutions.

____

____

____

____

____

____

____

____

Quick Lab

Observing Osmosis

You can observe the movement of water into or out of a grape under different conditions.

MATERIALS

- 3 grapes
- 3 small jars with lids
- saturated sugar solution
- grape juice
- tap water
- marking pen
- paper towel
- balance

Procedure

1. You will be using the data table below to record your data.

Data Table			
Solution	**Original Mass**	**Predicted Mass**	**Actual Mass**
Sugar solution			
Grape juice			
Water			

2. Fill one jar with the sugar solution. Fill a second jar with grape juice. (The grape will be more visible inside the jar if you fill the jar with white grape juice.) Fill the third jar with tap water. Label each jar according to the solution it contains.

3. Using the balance, find the mass of each grape. Place one grape in each jar, and record the mass of each jar in the data table. Put a lid on each jar.

4. Predict whether the mass of each grape will increase or decrease over time. Explain your predictions.

5. After 24 hours, remove each grape from its jar, and dry it gently with a paper towel. Using the balance, find its mass again. Record your results.

6. Clean up your materials before leaving the lab.

Analysis

1. Identify the solutions in which osmosis occurred.

____

____

____

____

2. Critical Thinking
Evaluating Conclusions How did you determine whether osmosis occurred in each of the three solutions?

____

____

____

3. Critical Thinking
Evaluating Hypotheses Did the mass of each grape change as you predicted? Why or why not?

____

____

____

____

____

Data Lab   **DATASHEET FOR IN-TEXT LAB**

Analyzing the Effect of Electrical Charge on Ion Transport

Background

The electrical charge of an ion affects the diffusion of the ion across the cell membrane. Some ions are more concentrated inside cells, and some ions are more concentrated outside cells. Use the table to answer the following questions:

Ion Charges and Concentration Inside and Outside Cell		
Ion	**Charge of ion**	**Concentration of ion outside cell : inside cell**
Sodium (Na^+)	Positive	10:1
Potassium (K^+)	Positive	1:20
Calcium (Ca^{2+})	Positive	10,000:1
Chloride (Cl^-)	Negative	12:1

Analysis

1. Identify the ion that is more concentrated inside the cell than outside the cell.

____

2. Identify those ions that are more concentrated outside the cell than inside the cell.

____

3. Critical Thinking
Recognizing Relationships Do the positive charges of calcium ions and sodium ions make these ions more likely to move into or out of the cell?

____

____

4. Critical Thinking
Inferring Relationships Which ions' electrical charges oppose the direction of movement that is caused by their concentration gradient?

____

____

____

____

Exploration Lab)   **DATASHEET FOR IN-TEXT LAB**

Analyzing the Effect of Cell Size on Diffusion

SKILLS

- Using scientific methods
- Collecting, organizing, and graphing data

OBJECTIVES

- **Relate** the size of a cell to its surface area-to-volume ratio.
- **Predict** how the surface area-to-volume ratio of a cell will affect the diffusion of substances into the cell.

MATERIALS

- safety goggles
- lab apron
- disposable gloves
- block of phenolphthalein agar ($3 \times 3 \times 6$ cm)
- plastic knife
- metric ruler
- 250 mL beaker
- 150 mL of vinegar
- plastic spoon
- paper towel

CHEMSAFETY

CAUTION: Always wear safety goggles and a lab apron to protect your eyes and clothing.

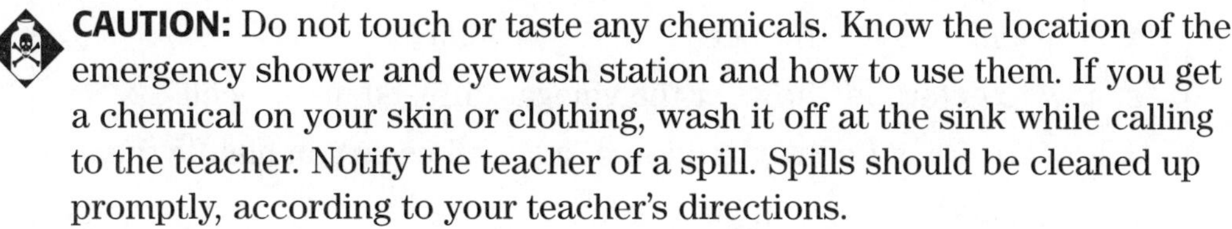

CAUTION: Do not touch or taste any chemicals. Know the location of the emergency shower and eyewash station and how to use them. If you get a chemical on your skin or clothing, wash it off at the sink while calling to the teacher. Notify the teacher of a spill. Spills should be cleaned up promptly, according to your teacher's directions.

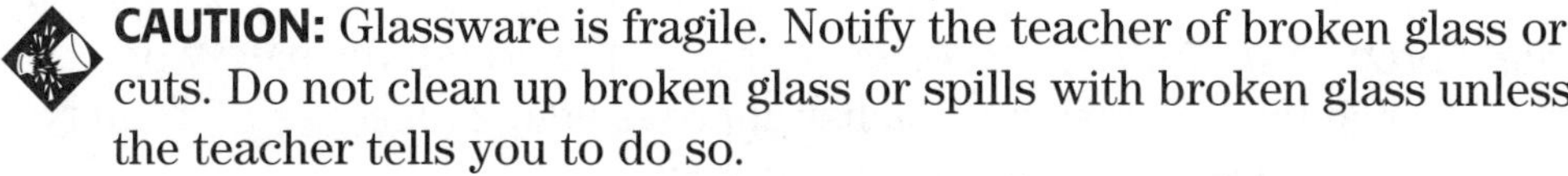

CAUTION: Glassware is fragile. Notify the teacher of broken glass or cuts. Do not clean up broken glass or spills with broken glass unless the teacher tells you to do so.

Before You Begin

Substances enter and leave a cell in several ways, including by **diffusion.** How efficiently a cell can exchange substances depends on the **surface area-to-volume ratio** (surface area ÷ volume) of the cell. **Surface area** is the size of the outside of an object. **Volume** is the amount of space an object takes up. In this lab, you will investigate how cell size affects the diffusion of substances into a cell. To do this, you will make cell models using agar that contains an indicator. This indicator will change color when an acidic solution diffuses into it.

1. Write a definition for each boldface term in the paragraph above. Use a separate sheet of paper.

2. Based on the objectives for this lab, write a question you would like to explore about cell size and diffusion.

____

____

Procedure

PART A: DESIGN AN EXPERIMENT

1. Work with members of your lab group to explore one of the questions written for step 2 of **Before You Begin.** To explore the question, design an experiment that uses the materials listed for this lab.

You Choose

As you design your experiment, decide the following:

  a. what question you will explore

  b. what hypothesis you will test

  c. how many "cells" (agar cubes) you will have and what sizes they will be

  d. how long to leave the "cells" in the vinegar

  e. how to determine how far the vinegar diffused into a "cell"

  f. how to prevent contamination of agar cubes as you handle them

  g. what data to record in your data table

---

2. Write a procedure for your experiment. Make a list of all the safety precautions you will take. Have your teacher approve your procedure and safety precautions before you begin the experiment.

PART B: CONDUCT YOUR EXPERIMENT

3. Put on safety goggles, a lab apron, and disposable gloves.

| Analyzing the Effect of Cell Size on Diffusion *continued*

4. Carry out the experiment you designed. Record your observations in your data table.

PART C: CLEANUP AND DISPOSAL

5. Dispose of solutions, broken glass, and agar in the designated waste containers. Do not pour chemicals down the drain or put lab materials in the trash unless your teacher tells you to do so.

6. Clean up your work area and all lab equipment. Return lab equipment to its proper place. Wash your hands thoroughly before you leave the lab and after you finish all work.

Analyze and Conclude

1. Summarizing Results Describe any changes in the appearance of the cubes.

____

____

2. Summarizing Results Make a graph using your group's data. Plot "Diffusion Distance (mm)" on the vertical axis. Use graph paper to plot "Surface Area-to-Volume Ratio" on the horizontal axis.

____

____

3. Analyzing Results Using the graph you made in item 2, make a statement about the relationship between the surface area-to-volume ratio and the distance a substance diffuses.

____

____

4. Summarizing Results Make a graph using your group's data. Use graph paper to plot "Rate of Diffusion (mm/min)" (distance vinegar moved ÷ time) on the vertical axis. Plot "Surface Area-to-Volume Ratio" on the horizontal axis.

____

____

5. Analyzing Results Using the graph you made in item 4, make a statement about the relationship between the surface area-to-volume ratio and the rate of diffusion of a substance.

____

____

Analyzing the Effect of Cell Size on Diffusion *continued*

6. Evaluating Methods In what ways do your agar models simplify or fail to simulate the features of real cells?

____

____

____

7. Calculating Calculate the surface area and volume of a cube with a side length of 5 cm. Calculate the surface area and volume of a cube with a side length of 10 cm. Determine the surface area-to-volume ratio of each of these cubes. Which cube has the greater surface area-to-volume ratio?

____

____

____

____

8. Evaluating Conclusions How does the size of a cell affect the diffusion of substances into the cell?

____

____

____

9. Further Inquiry Write a new question about cell size and diffusion that could be explored with another investigation.

____

____

Demonstrating Diffusion

Molecules must move about in a cell in order for a cell to survive. In this lab, you will see the movement of molecules and study how temperature affects this movement.

OBJECTIVES

Observe diffusion.

Recognize the relationship between temperature and the rate of diffusion.

MATERIALS

- beaker, at least 400 mL
- beakers, 250 mL (3)
- cornstarch
- gloves
- hot plate
- ice cubes
- lab apron
- Lugol's iodine solution
- oven mitt
- resealable plastic bags (3)
- safety goggles
- teaspoon
- thermometer
- timer
- water
- wax pencil

Procedure

1. Put on safety goggles, a lab apron, and gloves. Label each of three clean 250 mL beakers "cold," "medium," and "hot."

2. Add 90 mL of cool or lukewarm tap water to each beaker.

3. Place two ice cubes in the "cold" beaker. Set the beakers aside.

4. In a larger beaker, dissolve 1 teaspoon of cornstarch in 250 mL of water. Pour the mixture into three plastic bags, distributing it evenly among the bags. Seal the bags.

5. Add 10 mL of Lugol's iodine to the "medium" beaker. **CAUTION: Lugol's iodine is a poison and irritant. Avoid eye and skin contact. Do not inhale or ingest. This material also will stain skin and clothing.** Use a thermometer to find the temperature of the water in the beaker. Record this initial temperature in **Table 1.**

6. Place a bag with the cornstarch solution in the "medium" beaker. Observe what happens in the beaker. Record your observations in **Table 1.** Include notes about how long it took noticeable changes to occur.

Name _________________________________ Class _______________ Date _____________

7. Place the "hot" beaker on a hot plate, and heat the water to boiling. **CAUTION: A hot plate's high temperature can cause injury. Make sure the electrical cord is not in the way of your movements.**

TABLE 1 TEMPERATURE AND DIFFUSION TIMES

	Temperature of beaker		
	Cold	Medium	Hot
Initial temperature			
Observations			

8. Use an oven mitt to carefully remove the beaker from the hot plate. **CAUTION: Boiling water can cause injury.** Place the beaker on a surface that is suited for high temperatures.

9. Repeat steps 5 and 6 for the "hot" and "cold" beakers. Record your results in **Table 1.**

10. Clean up your materials, and wash your hands before leaving the lab.

Analysis and Conclusions

1. Describing Events What happened to the water inside the bag?

____

2. Explaining Events Did the cornstarch diffuse out of the bag? How do you know?

____

____

3. Drawing Conclusions Explain how temperature affects diffusion.

____

____

____

4. Making Inferences Why is a warm body advantageous for a living thing?

____

____

  DEMONSTRATION

Observing Osmosis in Eggs

Some chemicals can pass through a cell membrane, but others cannot. Furthermore, not all chemicals can pass through a cell membrane with equal ease. The cell membrane determines which chemicals can diffuse into or out of a cell.

As chemicals pass into and out of a cell, they move from areas of high concentration to areas of low concentration. Cells in *hypertonic* solutions have solute concentrations lower than the solution that bathes them. This concentration difference causes water to move out of the cell into the solution. Cells in *hypotonic* solutions have solute concentrations greater than the solution that bathes them. This concentration difference causes water to move from the solution into the cell. The movement of water into and out of a cell through the cell membrane is called *osmosis.*

In this lab, you will use a model of a living cell to predict the results of an experiment that involves the movement of water through a membrane.

OBJECTIVES

Explain changes that occur in a cell as a result of diffusion.

Distinguish between hypertonic and hypotonic solutions.

MATERIALS

- balance
- beakers, 250 mL (2)
- beakers, 600 mL (2)
- corn syrup
- distilled water
- eggs (2)

- lab apron
- paper towels (2)
- safety goggles
- tablespoon or tongs
- vinegar, 400 mL
- wax pencil

Procedure

DAY 1: SOAKING EGGS IN VINEGAR

1. Label one 600 mL beaker "Egg 1: water" and the other 600 mL beaker "Egg 2: syrup." Also label the beakers with the initials of each member of your group.

2. Measure the mass of each of two eggs to the nearest 0.1 g, and record your measurements in the second column of **Table 1. CAUTION: Uncooked eggs may contain harmful bacteria. Do not touch your face after you have handled raw eggs. Clean up any material from broken eggs immediately. Wash your hands with soap and water after handling the eggs.**

3. Put on safety goggles and a lab apron. Pour 200 mL of vinegar into each labeled beaker. Using a tablespoon or tongs, place an egg into each beaker. Always return each egg to the same beaker.

Observing Osmosis in Eggs *continued*

TABLE 1 EGGS IN VINEGAR

Egg	Mass of fresh egg with shell	Observations after 24 h	Mass after 24 h in vinegar
1			
2			

4. Place a 250 mL beaker containing 100 mL of water on each egg to keep it submerged, as shown in **Figure 1.** Add more vinegar if the egg is not covered by the vinegar already in the beaker. If some vinegar spills over when the 250 mL beaker is placed on the egg, carry the beaker carefully to the sink and pour out some vinegar. Store the beakers for 24 hours in the area specified by your teacher.

FIGURE 1 EGG IN VINEGAR

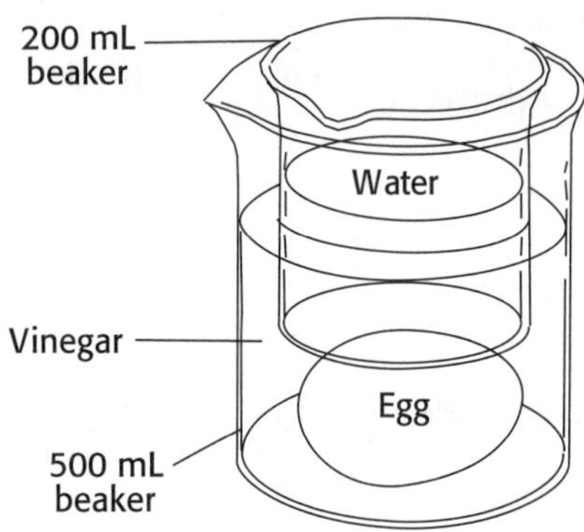

5. Clean up your work area and wash your hands before leaving the lab.

DAY 2: SOAKING EGGS IN TWO LIQUIDS

6. After 24 hours, observe the eggs. Record your observations in **Table 1.**

7. Put on safety goggles and a lab apron. Label two separate sheets of paper towel "Egg 1" and "Egg 2." Pour the vinegar from the beakers into the sink. Using a tablespoon or tongs, remove the eggs and rinse them with water. Place each egg on the appropriately labeled paper towel. Measure the mass of each egg, and record the measurement in the last column of **Table 1.**

8. Return Egg 1 to its beaker, and add water until the egg is covered. Return Egg 2 to its beaker, and add corn syrup until the egg is covered. Store the beakers for 24 hours in the same place as before.

9. Clean up your work area and wash your hands before leaving the lab.

Observing Osmosis in Eggs *continued*

DAY 3: MEASURING CHANGES IN THE EGGS

10. Predict how the mass of each egg has changed after 24 hours in each liquid. (Hint: An egg is surrounded by a membrane. Inside the membrane, the egg white consists mainly of water and dissolved protein. The yolk consists mainly of fat and water. Corn syrup is sugar dissolved in water. The protein, fat, and sugar are solutes.) Record your predictions in **Table 2.**

  • What will have occurred if your egg gains or loses mass?

  ____

11. Observe your eggs. Record your observations in **Table 2.** Measure and record the final masses of the two eggs.

TABLE 2 EGGS SOAKED IN TWO LIQUIDS

Egg	Liquid	Predicted change after 24 h	Observations after 24 h	Final Mass of egg
1				
2				

12. Dispose of your materials according to your teacher's instructions.

13. Clean up your work area, and wash your hands before leaving the lab.

Analysis

1. Describing Events What effect did the vinegar have on the eggs?

____

____

2. Explaining Events What caused the change in appearance in Egg 1 after it soaked in water?

____

3. Explaining Events What caused the mass of the egg to increase after soaking in the vinegar solution?

____

____

4. Analyzing Results What material seems to have moved through the membrane of Egg 2 after it soaked in the corn syrup? In what direction did the material move?

____

____

Conclusions

1. Evaluating Results How did your results in step 11 compare with your prediction?

____

____

2. Drawing Conclusions Which egg was in a hypertonic solution? Explain.

____

____

____

____

3. Drawing Conclusions Which egg was in a hypotonic solution? Explain.

____

____

____

____

4. Applying Conclusions What do you think would happen to a red blood cell placed in a test tube of distilled water? Explain.

____

____

____

Extensions

1. Designing Experiments Design an experiment to find out the effects of placing living yeast cells in hypertonic and hypotonic solutions.

2. Research and Communications Look up the chemical compositions of egg shell and of vinegar. Use this information to write a chemical equation that explains how egg shell dissolves in vinegar.

Name _________________________________ Class _________________ Date _____________

  DATASHEET FOR IN-TEXT LAB

Observing Osmosis

You can observe the movement of water into or out of a grape under different conditions.

MATERIALS

- 3 grapes
- 3 small jars with lids
- saturated sugar solution
- grape juice
- tap water
- marking pen
- paper towel
- balance

Procedure

1. You will be using the data table below to record your data.

Data Table			
Solution	**Original Mass**	**Predicted Mass**	**Actual Mass**
Sugar solution			
Grape juice			
Water			

2. Fill one jar with the sugar solution. Fill a second jar with grape juice. (The grape will be more visible inside the jar if you fill the jar with white grape juice.) Fill the third jar with tap water. Label each jar according to the solution it contains.

3. Using the balance, find the mass of each grape. Place one grape in each jar, and record the mass of each jar in the data table. Put a lid on each jar.

4. Predict whether the mass of each grape will increase or decrease over time. Explain your predictions.

5. After 24 hours, remove each grape from its jar, and dry it gently with a paper towel. Using the balance, find its mass again. Record your results.

6. Clean up your materials before leaving the lab.

Name _____________________________ Class _____________ Date _____________

Observing Osmosis *continued*

Analysis

1. **Identify** the solutions in which osmosis occurred.

   Answers will vary. The tap water should cause water to move into the grape.

   The sugar solution should cause water to move out of the grape. Results with

   the grape juice may vary depending on the sugar content of the juice.

   ____

   ____

2. **Critical Thinking**
   Evaluating Conclusions How did you determine whether osmosis occurred in
   each of the three solutions?

   Whether osmosis occurs can be determined by noting an increase or a

   decrease in the mass of the grapes.

   ____

   ____

3. **Critical Thinking**
   Evaluating Hypotheses Did the mass of each grape change as you predicted?
   Why or why not?

   Answers will vary. Students who had a clear understanding of osmosis before

   the experiment probably will not change their thinking.

   ____

   ____

   ____

   ____

Name _______________________________ Class _______________ Date ___________

Data Lab   **DATASHEET FOR IN-TEXT LAB**

Analyzing the Effect of Electrical Charge on Ion Transport

Background

The electrical charge of an ion affects the diffusion of the ion across the cell membrane. Some ions are more concentrated inside cells, and some ions are more concentrated outside cells. Use the table to answer the following questions:

Ion Charges and Concentration Inside and Outside Cell		
Ion	**Charge of ion**	**Concentration of ion outside cell : inside cell**
Sodium (Na^+)	Positive	10:1
Potassium (K^+)	Positive	1:20
Calcium (Ca^{2+})	Positive	10,000:1
Chloride (Cl^-)	Negative	12:1

Analysis

1. **Identify** the ion that is more concentrated inside the cell than outside the cell.

   potassium ion

2. **Identify** those ions that are more concentrated outside the cell than inside the cell.

   sodium ion, calcium ion, and chloride ion

3. **Critical Thinking**
   Recognizing Relationships Do the positive charges of calcium ions and sodium ions make these ions more likely to move into or out of the cell?

   Calcium ions and sodium ions are more likely to move into the cell because

   of their positive charges.

4. **Critical Thinking**
   Inferring Relationships Which ions' electrical charges oppose the direction of movement that is caused by their concentration gradient?

   Potassium ions move down their concentration gradient out of the cell.

   This movement, however, is opposed by the ions' attraction to the negatively

   charged interior of the cell. Chloride ions move down their concentration

   gradient into the cell, but this movement is opposed by the ions' being

   repelled by the negatively charged interior of the cell.

Name _______________________________ Class _______________ Date ___________

  DATASHEET FOR IN-TEXT LAB

Analyzing the Effect of Cell Size on Diffusion

SKILLS

- Using scientific methods
- Collecting, organizing, and graphing data

OBJECTIVES

- **Relate** the size of a cell to its surface area-to-volume ratio.
- **Predict** how the surface area-to-volume ratio of a cell will affect the diffusion of substances into the cell.

MATERIALS

- safety goggles
- lab apron
- disposable gloves
- block of phenolphthalein agar (3 × 3 × 6 cm)
- plastic knife
- metric ruler
- 250 mL beaker
- 150 mL of vinegar
- plastic spoon
- paper towel

CHEMSAFETY

CAUTION: Always wear safety goggles and a lab apron to protect your eyes and clothing.

CAUTION: Do not touch or taste any chemicals. Know the location of the emergency shower and eyewash station and how to use them. If you get a chemical on your skin or clothing, wash it off at the sink while calling to the teacher. Notify the teacher of a spill. Spills should be cleaned up promptly, according to your teacher's directions.

CAUTION: Glassware is fragile. Notify the teacher of broken glass or cuts. Do not clean up broken glass or spills with broken glass unless the teacher tells you to do so.

Name _______________________________ Class _______________ Date _____________

Analyzing the Effect of Cell Size on Diffusion *continued*

Before You Begin

Substances enter and leave a cell in several ways, including by **diffusion.** How efficiently a cell can exchange substances depends on the **surface area-to-volume ratio** (surface area ÷ volume) of the cell. **Surface area** is the size of the outside of an object. **Volume** is the amount of space an object takes up. In this lab, you will investigate how cell size affects the diffusion of substances into a cell. To do this, you will make cell models using agar that contains an indicator. This indicator will change color when an acidic solution diffuses into it.

1. Write a definition for each boldface term in the paragraph above. Use a separate sheet of paper. **Answers appear in the TE for this lab.**

2. Based on the objectives for this lab, write a question you would like to explore about cell size and diffusion.

   Answers will vary. For example: How does surface area-to-volume ratio

   affect the diffusion of a substance into a cell?

Procedure

PART A: DESIGN AN EXPERIMENT

1. Work with members of your lab group to explore one of the questions written for step 2 of **Before You Begin.** To explore the question, design an experiment that uses the materials listed for this lab.

You Choose

As you design your experiment, decide the following:

   a. what question you will explore

   b. what hypothesis you will test

   c. how many "cells" (agar cubes) you will have and what sizes they will be

   d. how long to leave the "cells" in the vinegar

   e. how to determine how far the vinegar diffused into a "cell"

   f. how to prevent contamination of agar cubes as you handle them

   g. what data to record in your data table

2. Write a procedure for your experiment. Make a list of all the safety precautions you will take. Have your teacher approve your procedure and safety precautions before you begin the experiment.

PART B: CONDUCT YOUR EXPERIMENT

3. Put on safety goggles, a lab apron, and disposable gloves.

Name _______________________________ Class _______________ Date _____________

Analyzing the Effect of Cell Size on Diffusion *continued*

4. Carry out the experiment you designed. Record your observations in your data table.

PART C: CLEANUP AND DISPOSAL

5. Dispose of solutions, broken glass, and agar in the designated waste containers. Do not pour chemicals down the drain or put lab materials in the trash unless your teacher tells you to do so.

6. Clean up your work area and all lab equipment. Return lab equipment to its proper place. Wash your hands thoroughly before you leave the lab and after you finish all work.

Analyze and Conclude

1. Summarizing Results Describe any changes in the appearance of the cubes.

Cubes should appear light pink near their surfaces.

2. Summarizing Results Make a graph using your group's data. Plot "Diffusion Distance (mm)" on the vertical axis. Use graph paper to plot "Surface Area-to-Volume Ratio" on the horizontal axis.

The line plotted on the graph should be horizontal.

3. Analyzing Results Using the graph you made in item 2, make a statement about the relationship between the surface area-to-volume ratio and the distance a substance diffuses.

The diffusion distance is the same regardless of the surface area-to-volume

ratio.

4. Summarizing Results Make a graph using your group's data. Use graph paper to plot "Rate of Diffusion (mm/min)" (distance vinegar moved ÷ time) on the vertical axis. Plot "Surface Area-to-Volume Ratio" on the horizontal axis.

The line plotted on the graph should be horizontal.

5. Analyzing Results Using the graph you made in item 4, make a statement about the relationship between the surface area-to-volume ratio and the rate of diffusion of a substance.

The rate of diffusion is constant regardless of the surface area-to-volume

ratio.

Name _________________________________ Class _______________ Date _____________

Analyzing the Effect of Cell Size on Diffusion *continued*

6. **Evaluating Methods** In what ways do your agar models simplify or fail to simulate the features of real cells?

Agar models ignore the selective permeability of cell membranes, the role of

membrane proteins in facilitated diffusion and active transport, and other

mechanisms of cell transport.

7. **Calculating** Calculate the surface area and volume of a cube with a side length of 5 cm. Calculate the surface area and volume of a cube with a side length of 10 cm. Determine the surface area-to-volume ratio of each of these cubes. Which cube has the greater surface area-to-volume ratio?

For a cube with a side length of 5 cm: surface area = 150 cm^2, volume =

125 cm^3; surface area-to-volume ratio = 6:5. For a cube with side length of

10 cm: surface area = 600 cm^2; volume = 1,000 cm^3; surface area-to-volume

ratio = 3:5. The small cube has the greater surface area-to-volume ratio.

8. **Evaluating Conclusions** How does the size of a cell affect the diffusion of substances into the cell?

In a small cell, substances do not have to diffuse as far to reach the center

of the cell.

9. **Further Inquiry** Write a new question about cell size and diffusion that could be explored with another investigation.

Answers will vary. For example: How can the cell membrane be modified to

take in materials more efficiently?

OBSERVATION

Demonstrating Diffusion

Teacher Notes

TIME REQUIRED 30 minutes

SKILLS ACQUIRED

Collecting data
Experimenting
Identifying patterns
Inferring
Interpreting
Organizing and analyzing data

RATINGS

Easy ← 1   2   3   4 → Hard

Teacher Prep–2
Student Setup–2
Concept Level–2
Cleanup–2

THE SCIENTIFIC METHOD

Make Observations Students observe how temperature affects diffusion.

Analyze the Results Analysis and Conclusions question 2 requires students to analyze their results.

Draw Conclusions Analysis and Conclusions question 3 asks students to draw conclusions from their data.

MATERIALS

Materials for this lab can be purchased from WARD'S. See the *Master Materials List* for ordering instructions.

Lugol's iodine (iodine-iodide solution) is available ready-made. However, you can make it yourself. **CAUTION: Wear safety goggles and a lab apron if you prepare Lugol's iodine. Iodine reacts with metal, skin, and many other substances.** To make Lugol's iodine, dissolve 1.0 g of potassium iodide, KI, in 15 mL of distilled water in a 250 mL beaker. Add 0.7 g of iodine, I_2, and stir until dissolved, adding no more than 50 mL of water, if desired, to facilitate dissolution. Dilute to 100 mL while stirring well. Dispense in amber-colored dropper bottles labeled "Lugol's iodine solution." Sunlight or strong lights can cause the solution to deteriorate.

Demonstrating Diffusion *continued*

SAFETY CAUTIONS

- Discuss all safety symbols and caution statements with students.

- Lugol's iodine is a poison and irritant. Avoid eye and skin contact; do not ingest. This material will stain skin and clothing. In case of contact: *Ingestion:* If swallowed, give one or two glasses of milk to drink. Follow by giving starch solution, flour or egg white. Contact a physician immediately. *Eye Contact:* Flush eyes, including under the eyelids, thoroughly under running water for 15 minutes. Get immediate medical attention. *Inhalation:* Remove to fresh air; get immediate medical attention. *Skin:* Flush thoroughly with water. Contact a physician if irritation develops.

DISPOSAL

- Combine all wastes containing Lugol's iodine. Place 250 mL in a 1 L beaker. Slowly add 0.1 M sodium thiosulfate, and mix until the solution is decolorized (fully reduced). Place a beaker of the decolorized solution in the sink. Run water to overflowing for 10 minutes, flushing to a sanitary sewer.

- Pour leftover cornstarch and water down the drain.

- Provide a separate container for the disposal of broken glass.

TIPS AND TRICKS

Preparation

This lab works best in groups of two students.

Demonstrate to students the color change that occurs when iodine comes in contact with starch.

To save time, you could prepare the plastic bags with cornstarch in advance.

Instead of having each group carry out the experiment under cold, medium, and hot conditions, you might assign groups either the cold, medium, or hot beaker to observe. Then have students pool their data for comparison.

Procedure

Introduce the lab with a discussion of the importance of the movement of molecules into, out of, and within the cell.

If hot plates are not available, water can be boiled in a teapot or coffeemaker. If the use of ice cubes is inconvenient, water can be chilled overnight in a refrigerator. A 9°C spread of temperatures from hot to cold should produce easily observed differences in the rate of diffusion.

Name _______________________________ Class ________________ Date ______________

  OBSERVATION

Demonstrating Diffusion

Molecules must move about in a cell in order for a cell to survive. In this lab, you will see the movement of molecules and study how temperature affects this movement.

OBJECTIVES

Observe diffusion.

Recognize the relationship between temperature and the rate of diffusion.

MATERIALS

- beaker, at least 400 mL
- beakers, 250 mL (3)
- cornstarch
- gloves
- hot plate
- ice cubes
- lab apron
- Lugol's iodine solution
- oven mitt
- resealable plastic bags (3)
- safety goggles
- teaspoon
- thermometer
- timer
- water
- wax pencil

Procedure

1. Put on safety goggles, a lab apron, and gloves. Label each of three clean 250 mL beakers "cold," "medium," and "hot."

2. Add 90 mL of cool or lukewarm tap water to each beaker.

3. Place two ice cubes in the "cold" beaker. Set the beakers aside.

4. In a larger beaker, dissolve 1 teaspoon of cornstarch in 250 mL of water. Pour the mixture into three plastic bags, distributing it evenly among the bags. Seal the bags.

5. Add 10 mL of Lugol's iodine to the "medium" beaker. **CAUTION: Lugol's iodine is a poison and irritant. Avoid eye and skin contact. Do not inhale or ingest. This material also will stain skin and clothing.** Use a thermometer to find the temperature of the water in the beaker. Record this initial temperature in **Table 1.**

6. Place a bag with the cornstarch solution in the "medium" beaker. Observe what happens in the beaker. Record your observations in **Table 1.** Include notes about how long it took noticeable changes to occur.

Name _________________________ Class ______________ Date ____________

Demonstrating Diffusion *continued*

7. Place the "hot" beaker on a hot plate, and heat the water to boiling.
CAUTION: A hot plate's high temperature can cause injury. Make sure the electrical cord is not in the way of your movements.

TABLE 1 TEMPERATURE AND DIFFUSION TIMES

	Temperature of beaker		
	Cold	Medium	Hot
Initial temperature	12.1° C	31.2° C	81.4° C
Observations	After 28 minutes, the solution is slightly blue.	After 8 minutes, the solution is bluish-purple.	After 4 minutes the solution is bluish-black.

8. Use an oven mitt to carefully remove the beaker from the hot plate. **CAUTION: Boiling water can cause injury.** Place the beaker on a surface that is suited for high temperatures.

9. Repeat steps 5 and 6 for the "hot" and "cold" beakers. Record your results in **Table 1.**

10. Clean up your materials, and wash your hands before leaving the lab.

Analysis and Conclusions

1. Describing Events What happened to the water inside the bag?

The water in the bag turned black or blue as the iodine diffused into the bag.

2. Explaining Events Did the cornstarch diffuse out of the bag? How do you know?

The cornstarch did not diffuse out of the bag. The iodine in the beaker would

have turned black or blue.

3. Drawing Conclusions Explain how temperature affects diffusion.

Answers will vary according to the data but should indicate that the rate of

diffusion is proportional to the temperature of the environment.

4. Making Inferences Why is a warm body advantageous for a living thing?

Students should understand that warm body temperatures increase the rate

of diffusion without the cell expending any energy.

Skills Practice Lab)   **DEMONSTRATION**

Observing Osmosis in Eggs

Teacher Notes

TIME REQUIRED Three 25-minute periods

SKILLS ACQUIRED
Collecting data
Predicting
Inferring
Interpreting
Measuring
Organizing and analyzing data

RATINGS   Easy ◄—1——2——3——4—► Hard
Teacher Prep–2
Student Setup–2
Concept Level–2
Cleanup–2

THE SCIENTIFIC METHOD

Make Observations Procedure steps 6 and 11 require students to make and record observations. Analysis question 1 requires students to report their observations.

Analyze the Results Analysis questions 2–4 require students to analyze their results.

Draw Conclusions Conclusions questions 2 and 3 ask students to draw conclusions from their data.

MATERIALS

Materials for this lab can be purchased from WARD'S. See the *Master Materials List* for ordering instructions.

SAFETY CAUTIONS

• Discuss all safety symbols and caution statements with students.

• To avoid possible salmonella poisoning from raw eggs, be sure that students avoid touching their faces during this lab. Also ensure that students thoroughly wash their hands and clean the work surfaces after each day's procedures. Have students carefully wash all glassware after use with the eggs.

DISPOSAL

• Dispose of the used eggs in a leak-resistant garbage bag to prevent contamination.

• Solutions may be flushed down the drain with plenty of water.

Observing Osmosis in Eggs *continued*

TIPS AND TRICKS

Preparation

This lab works best in groups of two to four students. To save time and materials, the lab can be done as a demonstration.

Review the terms *solute* and *solution* with students. A solution is a homogeneous mixture of two or more substances. The substance present in the greater amount (usually a liquid) is the solvent. The substance present in the lesser amount is the solute. Explain that vinegar is a solution.

Procedure

Students will need three consecutive days to perform this lab. However, the procedure for each day will not require an entire lab period. To shorten the lab to two days, have students begin with eggs that have already soaked in vinegar for 24 hours and compare the soaked eggs with a sample raw egg. Students should begin at Procedure step 6.

In step 8, consider adding food coloring to the water to which students return Egg 1. At the end of the lab, have students pierce the end of the egg. The colored water will come out, dramatically illustrating that the water entered the egg through osmosis.

If students cannot perform this lab on consecutive days, the eggs can be left in their respective solutions at room temperature for several days with no ill effects.

Name _________________________________ Class _______________ Date _____________

Skills Practice Lab)   **DEMONSTRATION**

Observing Osmosis in Eggs

Some chemicals can pass through a cell membrane, but others cannot. Furthermore, not all chemicals can pass through a cell membrane with equal ease. The cell membrane determines which chemicals can diffuse into or out of a cell.

As chemicals pass into and out of a cell, they move from areas of high concentration to areas of low concentration. Cells in *hypertonic* solutions have solute concentrations lower than the solution that bathes them. This concentration difference causes water to move out of the cell into the solution. Cells in *hypotonic* solutions have solute concentrations greater than the solution that bathes them. This concentration difference causes water to move from the solution into the cell. The movement of water into and out of a cell through the cell membrane is called *osmosis.*

In this lab, you will use a model of a living cell to predict the results of an experiment that involves the movement of water through a membrane.

OBJECTIVES

Explain changes that occur in a cell as a result of diffusion.

Distinguish between hypertonic and hypotonic solutions.

MATERIALS

- balance
- beakers, 250 mL (2)
- beakers, 600 mL (2)
- corn syrup
- distilled water
- eggs (2)

- lab apron
- paper towels (2)
- safety goggles
- tablespoon or tongs
- vinegar, 400 mL
- wax pencil

Procedure
DAY 1: SOAKING EGGS IN VINEGAR

1. Label one 600 mL beaker "Egg 1: water" and the other 600 mL beaker "Egg 2: syrup." Also label the beakers with the initials of each member of your group.

2. Measure the mass of each of two eggs to the nearest 0.1 g, and record your measurements in the second column of **Table 1. CAUTION: Uncooked eggs may contain harmful bacteria. Do not touch your face after you have handled raw eggs. Clean up any material from broken eggs immediately. Wash your hands with soap and water after handling the eggs.**

3. Put on safety goggles and a lab apron. Pour 200 mL of vinegar into each labeled beaker. Using a tablespoon or tongs, place an egg into each beaker. Always return each egg to the same beaker.

                              58

Name _________________________________ Class _______________ Date _______________

Observing Osmosis in Eggs *continued*

TABLE 1 EGGS IN VINEGAR

Egg	Mass of fresh egg with shell	Observations after 24 h	Mass after 24 h in vinegar
1	X	Shell is dissolved.	>X
2	X	Shell is dissolved.	>X

4. Place a 250 mL beaker containing 100 mL of water on each egg to keep it submerged, as shown in **Figure 1.** Add more vinegar if the egg is not covered by the vinegar already in the beaker. If some vinegar spills over when the 250 mL beaker is placed on the egg, carry the beaker carefully to the sink and pour out some vinegar. Store the beakers for 24 hours in the area specified by your teacher.

FIGURE 1 EGG IN VINEGAR

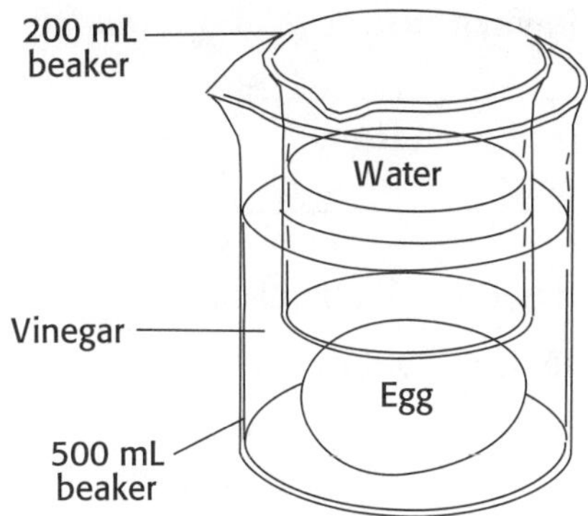

5. Clean up your work area and wash your hands before leaving the lab.

DAY 2: SOAKING EGGS IN TWO LIQUIDS

6. After 24 hours, observe the eggs. Record your observations in **Table 1.**

7. Put on safety goggles and a lab apron. Label two separate sheets of paper towel "Egg 1" and "Egg 2." Pour the vinegar from the beakers into the sink. Using a tablespoon or tongs, remove the eggs and rinse them with water. Place each egg on the appropriately labeled paper towel. Measure the mass of each egg, and record the measurement in the last column of **Table 1.**

8. Return Egg 1 to its beaker, and add water until the egg is covered. Return Egg 2 to its beaker, and add corn syrup until the egg is covered. Store the beakers for 24 hours in the same place as before.

9. Clean up your work area and wash your hands before leaving the lab.

Name _______________________________ Class _______________ Date _____________

Observing Osmosis in Eggs *continued*

DAY 3: MEASURING CHANGES IN THE EGGS

10. Predict how the mass of each egg has changed after 24 hours in each liquid. (Hint: An egg is surrounded by a membrane. Inside the membrane, the egg white consists mainly of water and dissolved protein. The yolk consists mainly of fat and water. Corn syrup is sugar dissolved in water. The protein, fat, and sugar are solutes.) Record your predictions in **Table 2.**

- What will have occurred if your egg gains or loses mass?

    Diffusion (osmosis) will have occurred.

11. Observe your eggs. Record your observations in **Table 2.** Measure and record the final masses of the two eggs.

TABLE 2 EGGS SOAKED IN TWO LIQUIDS

Egg	Liquid	Predicted change after 24 h	Observations after 24 h	Final Mass of egg
1	water	Predictions will vary.	Egg is very full.	increased
2	syrup	Predictions will vary.	Egg looks smaller. Membrane is looser.	decreased

12. Dispose of your materials according to your teacher's instructions.

13. Clean up your work area, and wash your hands before leaving the lab.

Analysis

1. Describing Events What effect did the vinegar have on the eggs?

The vinegar dissolved the shells.

2. Explaining Events What caused the change in appearance in Egg 1 after it soaked in water?

Water moved through the membrane into the egg. Osmosis occurred.

3. Explaining Events What caused the mass of the egg to increase after soaking in the vinegar solution?

Water moved from the vinegar into the egg.

Name _______________________________ Class _______________ Date _______________

Observing Osmosis in Eggs *continued*

4. Analyzing Results What material seems to have moved through the membrane of Egg 2 after it soaked in the corn syrup? In what direction did the material move?

Water moved through the membrane out of the egg.

Conclusions

1. Evaluating Results How did your results in step 11 compare with your prediction?

Answers will vary, depending on students' predictions.

2. Drawing Conclusions Which egg was in a hypertonic solution? Explain.

Egg 2 was in a hypertonic solution because there was a higher solute concentration in the syrup than in the egg. As a result, water diffused from the egg into the syrup, decreasing the egg's mass.

3. Drawing Conclusions Which egg was in a hypotonic solution? Explain.

Egg 1 was in a hypotonic solution because there was a higher solute concentration in the egg than in the water. As a result, water diffused into the egg, increasing the egg's mass.

4. Applying Conclusions What do you think would happen to a red blood cell placed in a test tube of distilled water? Explain.

The distilled water is hypotonic. Thus, water would diffuse into the red blood cell, causing it to swell and possibly burst.

Extensions

1. Designing Experiments Design an experiment to find out the effects of placing living yeast cells in hypertonic and hypotonic solutions.

2. Research and Communications Look up the chemical compositions of egg shell and of vinegar. Use this information to write a chemical equation that explains how egg shell dissolves in vinegar.

Answer Key

Directed Reading

SECTION: PASSIVE TRANSPORT

1. Passive transport is movement across a cell membrane that does not require energy from the cell. Osmosis is an example of passive transport because it does not require energy from the cell.
2. Cell membranes control the movement of substances into and out of cells.
3. Substances diffuse across the cell membrane from an area of high concentration to an area of lower concentration.
4. Equilibrium is a condition in which the concentration of a substance is equal throughout a space.
5. Osmosis is the diffusion of water through a selectively permeable membrane. Diffusion is the movement of a substance from an area of high concentration to an area of lower concentration.
6. A hypertonic solution causes a cell to shrink as water moves out of the cell by osmosis. A hypotonic solution causes a cell to swell as water moves into the cell by osmosis.
7. An isotonic solution has no effect on cell volume. In a solution, equilibrium is a state in which there is no net movement of substances. Cells are in a state of equilibrium in an isotonic solution.
8. f
9. e
10. c
11. b
12. g
13. a
14. d

SECTION: ACTIVE TRANSPORT

1. active transport
2. energy
3. ATP
4. carrier
5. less
6. greater
7. potassium
8. potassium
9. Proteins and polysaccharides are too large.
10. Endocytosis involves the movement of substances into cells. Exocytosis involves the movement of substances out of cells.
11. In endocytosis, a cell membrane forms a pouch around a substance. The pouch closes up and pinches off from the membrane, enclosing the substance in a vesicle inside the cell.
12. In exocytosis, a vesicle fuses with the inner surface of a cell membrane. The outer surface of the membrane opens, releasing the contents of the vesicle outside the cell.
13. They keep the sodium content of the cell at a low level. Too much sodium in a cell would cause water to enter the cell by osmosis, causing the cell to swell or burst. They also maintain the concentration gradients of sodium and potassium ions, which cells use to transport substances such as glucose across cell membranes.
14. f
15. c
16. a
17. b
18. d
19. e
20. g

Active Reading

SECTION: PASSIVE TRANSPORT

1. osmosis; the diffusion of water through a selectively permeable membrane
2. Osmosis is a type of diffusion.
3. Other forms of diffusion involve movement of different substances down a concentration gradient.
4. d

SECTION: ACTIVE TRANSPORT

1. movement into cell
2. Cell membrane forms pouch around substances outside of cell or; vesicles may fuse with lysosomes or other organelles.
3. movement out of cell
4. Cell membrane fuses with vesicles in the cell; cells export proteins that are modified by the Golgi apparatus
5. The prefix indicates that something is occurring inside or within an object. In this case, a vesicle is moving a substance inside, or within a cell.
6. b

Vocabulary Review

1. h	**10.** e
2. j	**11.** m
3. i	**12.** o
4. a	**13.** d
5. k	**14.** g
6. b	**15.** c
7. l	**16.** q
8. n	**17.** p
9. f	

Science Skills

PREDICTING

1. Thirty grams of substance A will be in solution on side 1, and 30 g will be in solution on side 2. Because the membrane is permeable to substance A, diffusion will take place. Molecules of substance A will move from the side of the membrane where they are less concentrated until equilibrium is reached.
2. The original 20 g of substance B will remain in solution on side 1, and the original 40 g will remain in solution on side 2 because substance B cannot pass through the membrane.
3. The water level will be higher on side 2 of the U-tube. As substance A molecules move across the membrane from side 1 to side 2, there will be more dissolved molecules on side 2 than on side 1. This will cause water to move across the membrane by osmosis from side 1 to side 2.

4. The cells will swell and will take in water until they and the flask solution are isotonic.
5. The cells will remain the same size because the solutions are isotonic.
6. The cells will shrink, and water will move out of the cells until they and the flask solution are isotonic.

Concept Mapping

1. passive transport
2. concentration gradients
3. facilitated diffusion or ion channels
4. ion channels or facilitated diffusion
5. active transport
6. endocytosis
7. sodium-potassium pump or receptor proteins
8. receptor proteins or sodium-potassium pump

Critical Thinking

1. c	**12.** b
2. a	**13.** j, b
3. b	**14.** c, f
4. d	**15.** h, g
5. e	**16.** a, e
6. b	**17.** l, k
7. c	**18.** d, i
8. a	**19.** d
9. a	**20.** b
10. d	**21.** d
11. c	

Test Prep Pretest

1. c
2. d
3. b
4. a
5. a
6. endocytosis
7. communicate
8. sodium-potassium pump
9. against
10. facilitated diffusion
11. second messenger
12. gated ion channel
13. hypertonic
14. hypotonic
15. isotonic

16. The inside of a typical cell has a charge that is slightly negative; the outside of the cell is slightly positive. Because opposite charges attract, positively charged ions outside the cell are more likely to diffuse into the cell through ion channels in the membrane. Negatively charged ions are more likely to diffuse out of the cell through ion channels in the membrane. However, the rate of diffusion of these ions is also affected by their concentration gradient.

17. Answers will vary but may include the way that odors spread throughout a room.

18. The salt dissolves in the moisture on the surface of the pork chop. The resulting salt solution is far more concentrated than the fluid within the cells. Through osmosis, water molecules move from the cells to the surface of the pork chop, where water evaporates as the pork chop cooks. As this moisture from the interior evaporates, the pork chop will become drier.

19. Facilitated diffusion involves the movement of particles through carrier proteins.

20. The cell consumes food particles that are too big to pass through a protein channel by endocytosis.

Quiz

SECTION: PASSIVE TRANSPORT

1. c	6. c
2. d	7. b
3. b	8. d
4. a	9. a
5. c	10. d

SECTION: ACTIVE TRANSPORT

1. c	6. b
2. a	7. a
3. d	8. d
4. e	9. c
5. b	10. b

Chapter Test (General)

1. e	11. c
2. d	12. d
3. h	13. b
4. f	14. c
5. a	15. d
6. b	16. d
7. c	17. b
8. g	18. d
9. d	19. b
10. b	20. c

Chapter Test (Advanced)

1. c
2. a
3. b
4. c
5. d
6. d
7. c
8. e
9. b
10. a
11. energy
12. passive transport
13. equilibrium
14. negative
15. selectively
16. active transport
17. carrier proteins
18. endocytosis
19. receptor protein
20. hypertonic
21. Drinking sea water increases the concentration of salts in body fluids. This causes water to leave cells by osmosis, resulting in dehydration.
22. Using energy from ATP, the sodium-potassium pump transports three sodium ions out of the cell. The pump then picks up two potassium ions from outside the cell and transports them into the cell. The ions move against their concentration gradient. The process is important because it prevents a buildup of sodium in the cell, which can be toxic, and because it helps maintain the concentration gradients of sodium and potassium ions across the cell membrane.

23. (1) The receptor protein can act as an enzyme, chemically changing the molecules in the cytoplasm by activating other enzymes inside the cell. (2) The receptor protein may cause the formation of a second messenger that can change the functioning of the cell by activating enzymes or opening ion channels, for example. (3) The receptor protein can cause channels in the membrane to open, thereby changing the permeability of the cell membrane.

24. Diffusion is the random movement of a substance down its concentration gradient and does not require energy. Active transport uses energy to move particles against their concentration gradient. Active transport only occurs in cells, whereas diffusion can occur anywhere that a concentration gradient exists.

25. The rigid cell walls of plant and fungal cells prevent them from expanding too much as the cells take in water through osmosis. Some unicellular eukaryotes have contractile vacuoles that collect excess water and force it out of the cell. Many animal cells increase the "free" water concentration inside the cell by removing dissolved particles from the cytoplasm to maintain homeostasis. If cells are not able to prevent excess water from entering the cell, they may expand and eventually burst.

Lesson Plan

Section: Passive Transport

Pacing

Regular Schedule:	**with lab(s):** N/A	**without lab(s):** 3 days
Block Schedule:	**with lab(s):** N/A	**without lab(s):** 1 1/2 days

Objectives

1. Relate concentration gradients, diffusion, and equilibrium.

2. Predict the direction of water movement into and out of cells.

3. Describe the importance of ion channels in passive transport.

4. Identify the role of carrier proteins in facilitated diffusion.

National Science Education Standards Covered

UNIFYING CONCEPTS AND PROCESSES

UCP5: Form and function

SCIENCE AS INQUIRY

SI1: Abilities necessary to do scientific inquiry

SI2: Understandings about scientific inquiry

SCIENCE AND TECHNOLOGY

ST2: Understandings about science and technology

HISTORY AND NATURE OF SCIENCE

HNS2: Nature of scientific knowledge

LIFE SCIENCE: THE CELL

LSCell1: Cells have particular structures that underlie their functions.

LSCell2: Most cell functions involve chemical reactions.

LSCell3: Cells store and use information to guide their functions.

LSCell4: Cell functions are regulated.

PHYSICAL SCIENCE

PS 1: Structure of atoms

PS 2: Structure and properties of matter

PS 3: Chemical reactions

PS 4: Motions and forces

KEY
SE = Student Edition   TE = Teacher Edition
CRF = Chapter Resource File

Block 1
CHAPTER OPENER *(45 minutes)*

_ **Quick Review,** SE. Students answer questions covered in previous sections of the textbook as preparation for the chapter content. **(GENERAL)**

_ **Reading Activity,** SE. Students study the figures and captions in each section. For each figure, students write a question that can be answered by referring to the figure and its caption. **(GENERAL)**

_ **Using the Figure,** TE. Students answer questions about the chapter opener photograph. **(GENERAL)**

_ **Opening Activity,** TE. Students work together to construct a model of the cell membrane. **(GENERAL)**

Block 2
FOCUS *(5 minutes)*

_ **Bellringer Transparency.** Use this transparency as students enter the classroom and find their seats. **(GENERAL)**

MOTIVATE *(10 minutes)*

_ **Discussion/Question,** TE. Students describe at least one thing they have taught a pet to do and describe in detail the procedure they used to teach the pet. Then they explain why they used one particular method over another. **(BASIC)**

TEACH *(30 minutes)*

_ **Teaching Transparency, Section Outline.** Use this transparency to give students a framework for the information in this section. **(GENERAL)**

_ **Teaching Transparency, Osmosis.** Use this transparency to discuss the process of osmosis as a type of diffusion and passive transport. **(GENERAL)**

_ **Quick Lab,** Observing Osmosis, SE. Students the movement of water into and out of grapes under different conditions. Students setup up the lab on day 1 and then check their results the following day. **(GENERAL)**

_ **Datasheets for In-Text Labs,** Observing Osmosis, CRF.

_ **Teaching Transparency, Hypertonic, Hypotonic, and Isotonic Solutions.** Use this transparency to summarize the the effects of hypertonic, hypotonic, and isotonic solutions on cells. Ask students why it is important for the correct solute concentration to be maintained in the fluid that bathes cells. **(GENERAL)**

HOMEWORK

_ **Active Reading Worksheet, Passive Transport, CRF.** Students read a passage related to the section topic and answer questions. **(GENERAL)**

_ **Directed Reading Worksheet, Passive Transport, CRF.** Students complete the exercises in this worksheet to help them understand the material as they read the section. **(BASIC)**

Block 3

TEACH *(40 minutes)*

_ **Teaching Transparency, Ion Channels.** Use this transparency to discuss how ion channels allow certain ions to pass through the cell membrane. **(GENERAL)**

_ **Data Lab,** Analyzing the Effects of Electrical Charge on Ion Transport, SE. Students analyze data on ion charges and concentrations inside and outside of cells to answer questions. **(GENERAL)**

_ **Datasheets for In-Text Labs, Analyzing the Effects of Electrical Charge on Ion Transport, CRF.**

_ **Teaching Transparency, Facilitated Diffusion.** Use this transparency to discuss how facilitated diffusion works without using the cell's energy. **(GENERAL)**

_ **Quick Lab, Demonstrating Diffusion, CRF.** Students see the movement of molecules and study how temperature affects this movement. **(BASIC)**

CLOSE *(5 minutes)*

_ **Reteaching,** TE. Students work together to answer three questions provided in the TE. **(BASIC)**

HOMEWORK

_ **Quiz,** TE. Students answer questions that review the section material. **(GENERAL)**

_ **Section Review,** SE. Assign questions 1–5 for review, homework, or quiz. **(GENERAL)**

_ **Alternative Assessment**, TE. Students write a paragraph describing how the door on a house is like a cell membrane. **(GENERAL)**

_ **Quiz, CRF.** This quiz consists of ten multiple choice and matching questions that review the section's main concepts. **(BASIC) Also in Spanish.**

Other Resource Options

_ **Occupational Application Worksheet, Pharmacist, One-Stop Planner.** Students learn what a pharmacist does by filling out several patient profiles and completing a drug inventory form. **(GENERAL)**

- **Supplemental Reading Guide, The Lives of a Cell, One-Stop Planner.** Students read the book and answer questions. **(ADVANCED)**

- **Internet Connect.** Students can research Internet sources about Cell Membrane with SciLinks Code HX4035.

- **Internet Connect.** Students can research Internet sources about Water Movement in Cells with SciLinks Code HX4189.

- **Internet Connect.** Students can research Internet sources about Ion Channels with SciLinks Code HX4106.

- **go.hrw.com.** For worksheets, videos, and other teaching aids related to this chapter, visit the HRW Web site and type in the keyword HX4 ABH.

- **Biology Interactive Tutor CD-ROM,** Unit 1 Cell Transport and Homeostasis. Students watch animations and other visuals as the tutor explains cell transport and homeostasis. Students assess their learning with interactive activities.

- **CNN Science in the News, Video Segment 2 Cystic Fibrosis.** This video segment is accompanied by a **Critical Thinking Worksheet**.

- **CNN Student News.** Find the latest news, lesson plans, and activities related to important scientific events at **cnnstudentnews.com**.

Lesson Plan

Section: Active Transport

Pacing

Regular Schedule:	**with lab(s):** 3 days	**without lab(s):** 2 days
Block Schedule:	**with lab(s):** 1 1/2 days	**without lab(s):** 1 day

Objectives

1. Compare active transport with passive transport.

2. Describe the importance of the sodium-potassium pump.

3. Distinguish between endocytosis and exocytosis.

4. Identify three ways that receptor proteins can change the activity of a cell.

National Science Education Standards Covered

UNIFYING CONCEPTS AND PROCESSES

UCP5: Form and function

SCIENCE AS INQUIRY

SI1: Abilities necessary to do scientific inquiry

SI2: Understandings about scientific inquiry

SCIENCE AND TECHNOLOGY

ST2: Understandings about science and technology

HISTORY AND NATURE OF SCIENCE

HNS2: Nature of scientific knowledge

LIFE SCIENCE: THE CELL

LSCell1: Cells have particular structures that underlie their functions.

LSCell2: Most cell functions involve chemical reactions.

LSCell3: Cells store and use information to guide their functions.

LSCell4: Cell functions are regulated.

PHYSICAL SCIENCE

PS 2: Structure and properties of matter

PS 3: Chemical reactions

PS 4: Motions and forces

KEY
SE = Student Edition   TE = Teacher Edition
CRF = Chapter Resource File

Block 4

FOCUS *(5 minutes)*

_ **Bellringer Transparency.** Use this transparency as students enter the classroom and find their seats. **(GENERAL)**

MOTIVATE *(10 minutes)*

_ **Demonstration**, TE. A volunteer inflates a ball with an air pump. Then relate the activity toe a cell membrane pump. **(BASIC)**

TEACH *(30 minutes)*

_ **Teaching Transparency, Section Outline.** Use this transparency to give students a framework for the information in this section. **(GENERAL)**

_ **Teaching Transparency, Sodium-Potassium Pump.** Use this transparency to walk students through the steps of the sodium-potassium pump. **(GENERAL)**

_ **Teaching Tip,** Hydrogen Ion Pump, TE. Students suggest where hydrogen ion pumps are in animal and plants cells. **(GENERAL)**

_ **Teaching Transparency, Endocytosis and Exocytosis.** Use this transparency to discuss endocytosis and exocytosis as mechanisms of transport. **(GENERAL)**

_ **Group Activity,** Good and Bad Cholesterol, TE. Students work in groups to research and write a report on the difference between good and bad cholesterol. **(ADVANCED)**

HOMEWORK

_ **Directed Reading Worksheet, Active Transport, CRF.** Students complete the exercises in this worksheet to help them understand the material as they read the section. **(BASIC)**

_ **Active Reading Worksheet, Active Transport, CRF.** Students read a passage related to the section topic and answer questions. **(GENERAL)**

Block 5

TEACH *(30 minutes)*

_ **Skill Builder,** Math Skills, TE. Students calculate the amount of glucose a person must consume in one day to meet their energy needs from glucose alone. (**ADVANCED**)

_ **Teaching Transparency, Changes in Permeability.** Use this transparency to lead students through the process shown. Point out that ion channels can be open or closed and this state affects the permeability of the membrane. (**GENERAL**)

_ **Teaching Transparency, Second Messengers.** Use this transparency to how a receptor protein can trigger the production of a second messenger inside the cell. (**GENERAL**)

_ **Real Life**, SE. Students research some medicines that bind to receptor proteins. (**GENERAL**)

CLOSE *(10 minutes)*

_ **Reteaching,** TE. Students write down what they know about different vocabulary terms from the chapter, then check their answers using the textbook. (**BASIC**)

_ **Quiz,** TE. Students answer questions that review the section material. (**GENERAL**)

HOMEWORK

_ **Alternative Assessment,** TE. Students produce a crosswork puzzle with clues using terms introduced in this section. (**GENERAL**)

_ **Section Review,** SE. Assign questions 1–6 for review, homework, or quiz. (**GENERAL**)

_ **Science Skills Worksheet,** CRF. Students read about two different experiments and then answer questions about how materials move across membranes in the experiments. (**GENERAL**)

_ **Quiz,** CRF. This quiz consists of ten multiple choice and matching questions that review the section's main concepts. (**BASIC**) **Also in Spanish.**

_ **Modified Worksheet, One-Stop Planner.** This worksheet has been specially modified to reach struggling students. (**BASIC**)

_ **Critical Thinking Worksheet,** CRF. Students answer analogy-based questions that review the section's main concepts and vocabulary. (**ADVANCED**)

Optional Block

LAB *(45 minutes)*

_ **Exploration Lab,** Analyzing the Effect of Cell Size on Diffusion, SE. Students make cell models and use them to investigate how cell size affects the diffusion of substances into a cell. (**GENERAL**)

_ **Datasheets for In-Text Labs, Analyzing the Effect of Cell Size on Diffusion, CRF.**

Other Resource Options

_ **Skills Practice Lab, Observing Osmosis in Eggs, CRF.** Students use eggs with a dissolved shell as a model for a living cell. Then they predict the results of an experiment that involves the movement of water through a membrane. (**GENERAL**)

_ **Supplemental Reading Guide, The Lives of a Cell, One-Stop Planner.** Students read the book and answer questions. (**ADVANCED**)

_ **Internet Connect.** Students can research Internet sources about Cell Membrane with SciLinks Code HX4035.

_ **Internet Connect.** Students can research Internet sources about Receptor Proteins with SciLinks Code HX4157.

_ **go.hrw.com.** For worksheets, videos, and other teaching aids related to this chapter, visit the HRW Web site and type in the keyword HX4 ABH.

_ **Biology Interactive Tutor CD-ROM,** Unit 1 Cell Transport and Homeostasis. Students watch animations and other visuals as the tutor explains cell transport and homeostasis. Students assess their learning with interactive activities.

_ **CNN Science in the News, Video Segment 2 Cystic Fibrosis.** This video segment is accompanied by a **Critical Thinking Worksheet**.

_ **CNN Student News.** Find the latest news, lesson plans, and activities related to important scientific events at **cnnstudentnews.com**.

Lesson Plan

End-of-Chapter Review and Assessment

Pacing

Regular Schedule: 2 days
Block Schedule: 1 day

KEY
SE = Student Edition   **TE** = Teacher Edition
CRF = Chapter Resource File

Block 6

REVIEW *(45 minutes)*

_ **Study Zone,** SE. Use the Study Zone to review the Key Concepts and Key Terms of the chapter and prepare students for the Performance Zone questions. **(GENERAL)**

_ **Performance Zone,** SE. Assign questions to review the material for this chapter. Use the assignment guide to customize review for sections covered. **(GENERAL)**

_ **Teaching Transparency, Concept Mapping.** Use this transparency to review the concept map for this chapter. **(GENERAL)**

Block 7

ASSESSMENT *(45 minutes)*

_ **Chapter Test, Cells and Their Environment, CRF.** This test contains 20 multiple choice and matching questions keyed to the chapter's objectives. **(GENERAL) Also in Spanish.**

_ **Chapter Test, Cells and Their Environment, CRF.** This test contains 25 questions of various formats, each keyed to the chapter's objectives. **(ADVANCED)**

_ **Modified Chapter Test, One-Stop Planner.** This test has been specially modified to reach struggling students. **(BASIC)**

Other Resource Options

_ **Vocabulary Review Worksheet, CRF.** Use this worksheet to review the chapter vocabulary. **(GENERAL) Also in Spanish.**

_ **Test Prep Pretest, CRF.** Use this pretest to review the main content of the chapter. Each question is keyed to a section objective. **(GENERAL) Also in Spanish.**

_ **Test Item Listing for ExamView® Test Generator, CRF.** Use the Test Item Listing to identify questions to use in a customized homework, quiz, or test.

_ **ExamView® Test Generator, One-Stop Planner.** Create a customized homework, quiz, or test using the HRW Test Generator program.

Cells and Their Environment

TRUE/FALSE

1. ____ During diffusion, molecules diffuse from a region where their concentration is low to a region where their concentration is higher, until the particles are evenly dispersed.
   Answer: False   Difficulty: I   Section: 1   Objective: 1

2. ____ When the concentration of dissolved particles outside a cell is equal to the concentration of dissolved particles inside the cell, the cell solution is isotonic.
   Answer: True   Difficulty: I   Section: 1   Objective: 1

3. ____ Membranes are selectively permeable if they allow only certain substances to move across them.
   Answer: True   Difficulty: I   Section: 1   Objective: 1

4. ____ A cell placed in a strong salt solution would probably burst because of osmosis.
   Answer: False   Difficulty: I   Section: 1   Objective: 2

5. ____ Water will diffuse out of a cell when the cell is placed in a hypertonic solution.
   Answer: True   Difficulty: I   Section: 1   Objective: 2

6. ____ Osmosis is the diffusion of starch molecules through a selectively permeable membrane.
   Answer: False   Difficulty: I   Section: 1   Objective: 2

7. ____ The binding of specific molecules to ion channels controls the ability of particular ions to cross the cell membrane.
   Answer: True   Difficulty: I   Section: 1   Objective: 3

8. ____ To pass through a cell membrane, water requires carrier proteins.
   Answer: False   Difficulty: I   Section: 1   Objective: 4

9. ____ In facilitated diffusion, carrier proteins require energy to transport substances across the cell membrane.
   Answer: False   Difficulty: I   Section: 1   Objective: 4

10. ____ The transport of specific particles down their concentration gradient through a membrane by carrier proteins is known as facilitated diffusion.
    Answer: True   Difficulty: I   Section: 1   Objective: 4

11. ____ Diffusion is an active process that requires a cell to expend a great deal of energy.
    Answer: False   Difficulty: I   Section: 1   Objective: 1

12. ____ Diffusion through ion channels is a form of active transport.
    Answer: False   Difficulty: I   Section: 2   Objective: 1

13. ____ Facilitated diffusion moves molecules and ions against their concentration gradient, while active transport moves molecules and ions down their concentration gradient.
    Answer: False   Difficulty: I   Section: 2   Objective: 1

14. ____ Passive transport uses ATP to move molecules against their concentration gradient.
    Answer: False   Difficulty: I   Section: 2   Objective: 1

15. ____ In active transport, energy is required to move a substance across a cell membrane.
    Answer: True   Difficulty: I   Section: 2   Objective: 1

16. ____ The sodium-potassium pump requires energy to move ions across the cell membrane.
    Answer: True   Difficulty: I   Section: 2   Objective: 2

17. ____ The sodium-potassium pump moves sodium ions and potassium ions against their concentration gradient.

    Answer: True   Difficulty: I   Section: 2   Objective: 2

18. ____ The sodium-potassium pump transports sodium ions out of a cell while causing potassium ions to move into the cell.

    Answer: True   Difficulty: I   Section: 2   Objective: 2

19. ____ The sodium-potassium pump uses ATP.

    Answer: True   Difficulty: I   Section: 2   Objective: 2

20. ____ Exocytosis is a process that uses vesicles to capture substances and bring them into a cell.

    Answer: False   Difficulty: I   Section: 2   Objective: 3

21. ____ Exocytosis helps the cell rid itself of wastes.

    Answer: True   Difficulty: I   Section: 2   Objective: 3

22. ____ During the process of exocytosis, the cell membrane extends to engulf substances that are too big to pass through the cell membrane.

    Answer: False   Difficulty: I   Section: 2   Objective: 3

23. ____ Exocytosis does not use energy to expel proteins from the cell.

    Answer: False   Difficulty: I   Section: 2   Objective: 3

24. ____ Receptor proteins pump sodium ions into a cell.

    Answer: False   Difficulty: I   Section: 2   Objective: 4

25. ____ Receptor proteins may cause the formation of a second messenger molecule inside a cell.

    Answer: True   Difficulty: I   Section: 2   Objective: 4

26. ____ A receptor protein sends signals into a cell by transporting a specific molecule through the cell membrane.

    Answer: False   Difficulty: I   Section: 2   Objective: 4

MULTIPLE CHOICE

27. As a result of diffusion, the concentration of many types of substances
    a. always remains greater inside a membrane.
    b. eventually becomes balanced on both sides of a membrane.
    c. always remains greater outside of a membrane.
    d. becomes imbalanced on both sides of a membrane.

    Answer: B   Difficulty: I   Section: 1   Objective: 1

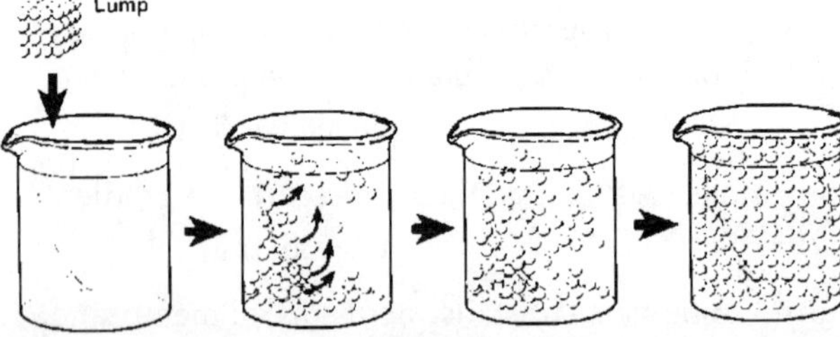

28. Refer to the illustration above. The process shown is called
    a. osmosis.   c. active transport.
    b. facilitated diffusion.   d. diffusion.

    Answer: D   Difficulty: II   Section: 1   Objective: 1

29. Diffusion is the movement of a substance
    a. only through a lipid bilayer membrane.
    b. from an area of low concentration to an area of higher concentration.
    c. only in liquids.
    d. from an area of high concentration to an area of lower concentration.

    Answer: D   Difficulty: I   Section: 1   Objective: 1

30. The dispersal of ink in a beaker of water is an example of
    a. diffusion.   c. active transport.
    b. osmosis.   d. endocytosis.

    Answer: A   Difficulty: I   Section: 1   Objective: 1

31. The diffusion of water into or out of a cell is called
    a. solubility.   c. selective transport.
    b. osmosis.   d. endocytosis.

    Answer: B   Difficulty: I   Section: 1   Objective: 2

32. Osmosis is a type of
    a. active transport.   c. facilitated diffusion.
    b. passive transport.   d. endocytosis.

    Answer: B   Difficulty: I   Section: 1   Objective: 2

33. A cell will swell when it is placed in a(n)
    a. hypotonic solution.   c. isotonic solution.
    b. hypertonic solution.   d. None of the above

    Answer: A   Difficulty: I   Section: 1   Objective: 2

34. The interior portion of a cell membrane forms a nonpolar zone that
    a. allows polar molecules to pass through the membrane.
    b. allows food to pass through the membrane.
    c. prevents ions and most large molecules from passing through the membrane.
    d. None of the above

    Answer: C   Difficulty: I   Section: 1   Objective: 3

35. Ions move through ion channels by
    a. endocytosis.   c. passive transport.
    b. diffusion.   d. active transport.

    Answer: C   Difficulty: I   Section: 1   Objective: 3

36. Ion channel gates close the pores of some ion channels in response to
    a. stretching of the cell membrane.
    b. a change in electrical charge.
    c. the binding of specific molecules to the channel.
    d. All of the above

    Answer: D   Difficulty: I   Section: 1   Objective: 3

37. Proteins that act like selective passageways in the cell membrane are known as
    a. marker proteins.   c. receptor proteins.
    b. channel proteins.   d. None of the above

    Answer: B   Difficulty: I   Section: 1   Objective: 3

38. Transport proteins that allow ions to pass through the cell membrane are called
    a. receptor proteins.   c. ion channels.
    b. marker proteins.   d. None of the above

    Answer: C   Difficulty: I   Section: 1   Objective: 3

39. Sugar molecules cross the cell membrane by
    a. active transport.
    b. facilitated diffusion.
    c. osmosis.
    d. gated channels.

    Answer: B   Difficulty: I   Section: 1   Objective: 4

40. Proteins involved in facilitated diffusion are
    a. carrier proteins.
    b. receptor proteins.
    c. Both (a) and (b)
    d. None of the above

    Answer: A   Difficulty: I   Section: 1   Objective: 4

41. Sugar molecules can enter cells through the process of
    a. exocytosis.
    b. facilitated diffusion.
    c. osmosis.
    d. ion pumps.

    Answer: B   Difficulty: I   Section: 1   Objective: 4

42. Which of the following does *not* require energy?
    a. diffusion
    b. endocytosis
    c. active transport
    d. sodium-potassium pump

    Answer: A   Difficulty: I   Section: 2   Objective: 1

43. Unlike passive transport, active transport
    a. requires energy.
    b. moves substances down their concentration gradient.
    c. does not involve carrier proteins.
    d. All of the above

    Answer: A   Difficulty: I   Section: 2   Objective: 1

44. Both active transport and facilitated diffusion involve
    a. ATP.
    b. movement against a concentration gradient.
    c. carrier proteins.
    d. All of the above

    Answer: C   Difficulty: I   Section: 2   Objective: 1

45. Which of the following is a form of active transport?
    a. osmosis
    b. diffusion
    c. facilitated diffusion
    d. sodium-potassium pump

    Answer: D   Difficulty: I   Section: 2   Objective: 1

46. The sodium-potassium pump
    a. is a carrier protein
    b. uses passive transport.
    c. is located in the cytoplasm of a cell.
    d. transports sugar molecules.

    Answer: A   Difficulty: I   Section: 2   Objective: 2

47. The sodium-potassium pump usually pumps
    a. potassium ions out of the cell.
    b. sodium ions into the cell.
    c. potassium ions into the cell.
    d. only potassium ions and sugar molecules.

    Answer: C   Difficulty: I   Section: 2   Objective: 2

48. The sodium-potassium pump
    a. increases the concentration of sodium ions inside a cell.
    b. decreases the concentration of sodium ions inside a cell.
    c. increases the concentration of potassium ions inside a cell.
    d. Both (b) and (c)

    Answer: D   Difficulty: I   Section: 2   Objective: 2

49. Proteins and polysaccharides that are too large to move into a cell through diffusion or active transport move in by
    a. exocytosis.
    b. endocytosis.
    c. the sodium-potassium pump.
    d. None of the above

    Answer: B   Difficulty: I   Section: 2   Objective: 3

50. Molecules that are too large to be moved through the membrane can be transported into the cell by
    a. osmosis.
    b. endocytosis.
    c. exocytosis.
    d. diffusion.

    Answer: B   Difficulty: I   Section: 2   Objective: 3

51. Molecules that are too large to be moved across a cell membrane can be removed from the cell by
    a. diffusion.
    b. exocytosis.
    c. endocytosis.
    d. osmosis.

    Answer: B   Difficulty: I   Section: 2   Objective: 3

52. Ridding the cell of materials by discharging the materials in vesicles is called
    a. osmosis.
    b. diffusion.
    c. exocytosis.
    d. endocytosis.

    Answer: C   Difficulty: I   Section: 2   Objective: 3

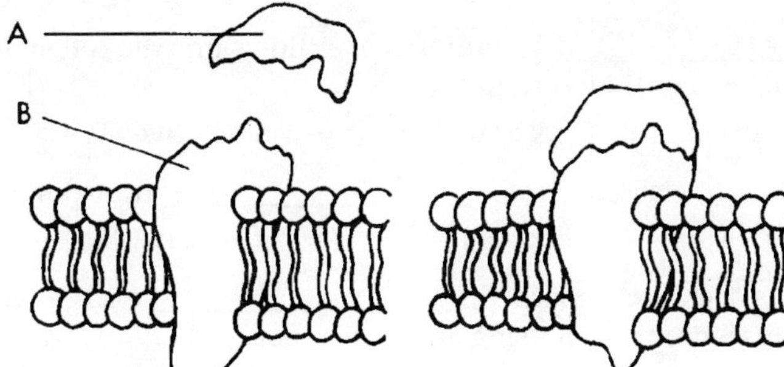

53. Refer to the illustration above. What happens when the structure labeled A binds to the structure labeled B?
    a. Information is sent into the cell.
    b. Proteins enter the cell.
    c. The cell begins to undergo mitosis.
    d. None of the above

    Answer: A   Difficulty: II   Section: 2   Objective: 4

54. Signal molecules bind to
    a. carbohydrates.
    b. marker proteins.
    c. receptor proteins.
    d. transport proteins.

    Answer: C   Difficulty: I   Section: 2   Objective: 4

55. When a signal molecule binds to a receptor protein, the receptor protein may
    a. change the permeability of the membrane.
    b. cause the formation of a second messenger molecule.
    c. catalyze certain chemical reactions in the cell.
    d. All of the above

    Answer: D   Difficulty: I   Section: 2   Objective: 4

56. Which of the following transmit information into a cell by binding to signal molecules?
    a. channel proteins
    b. receptor proteins
    c. marker proteins
    d. end proteins

    Answer: B   Difficulty: I   Section: 2   Objective: 4

COMPLETION

57. The random motion of particles of a substance that causes the substance to move from an area of high concentration to an area of lower concentration is called

____________________.

    Answer: diffusion   Difficulty: I   Section: 1   Objective: 1

58. The diffusion of ___________________ through cell membranes is called osmosis.

    Answer: water   Difficulty: I   Section: 1   Objective: 1

59. Substances always flow from an area of high concentration to an area of ___________________ concentration.

    Answer: low   Difficulty: I   Section: 1   Objective: 1

60. When the concentration of free water molecules is higher outside a cell than inside the cell, water will diffuse ___________________ the cell.

    Answer: into   Difficulty: I   Section: 1   Objective: 2

61. If a cell is placed in a(n) ___________________ solution, water will flow out of the cell.

    Answer: hypertonic   Difficulty: I   Section: 1   Objective: 2

62. If a cell is placed in a(n) ___________________ solution, water will flow into the cell.

    Answer: hypotonic   Difficulty: I   Section: 1   Objective: 2

63. If a cell is placed in a(n) ___________________ solution, water flows into the cell at a rate that is equal to the rate at which water flows out of the cell.

    Answer: isotonic   Difficulty: I   Section: 1   Objective: 2

64. Diffusion of ions through ion channels is a form of ___________________ transport.

    Answer: passive   Difficulty: I   Section: 1   Objective: 3

65. If the interior of a typical cell is negatively charged, ___________________ charged ions will not require energy to diffuse into the cell using an ion channel.

    Answer: positively   Difficulty: II   Section: 1   Objective: 3

66. In facilitated diffusion, ___________________ proteins are used to transport substances down their concentration gradient.

    Answer: carrier   Difficulty: I   Section: 1   Objective: 4

67. In ___________________ ___________________, carrier proteins do not require energy to transport amino acids into a cell.

    Answer: facilitated diffusion   Difficulty: I   Section: 1   Objective: 4

68. Carrier proteins ___________________ shape to transport sugars to the interior of cells.

    Answer: change   Difficulty: I   Section: 1   Objective: 4

69. A cell does not expend ___________________ when diffusion takes place.

    Answer: energy   Difficulty: I   Section: 2   Objective: 1

70. Active transport requires the use of ___________________ by a cell.

    Answer: ATP or energy   Difficulty: I   Section: 2   Objective: 1

71. The ___________________-___________________ pump transports ions against their concentration gradients.

    Answer: sodium potassium   Difficulty: I   Section: 2   Objective: 2

72. The sodium-potassium pump uses energy supplied by ___________________.

    Answer: ATP   Difficulty: I   Section: 2   Objective: 2

73. The sodium-potassium pump prevents the accumulation of _________________ ions inside the cell.

    Answer: sodium   Difficulty: I   Section: 2   Objective: 2

74. The movement of a substance into a cell by a vesicle is called _________________.

    Answer: endocytosis   Difficulty: I   Section: 2   Objective: 3

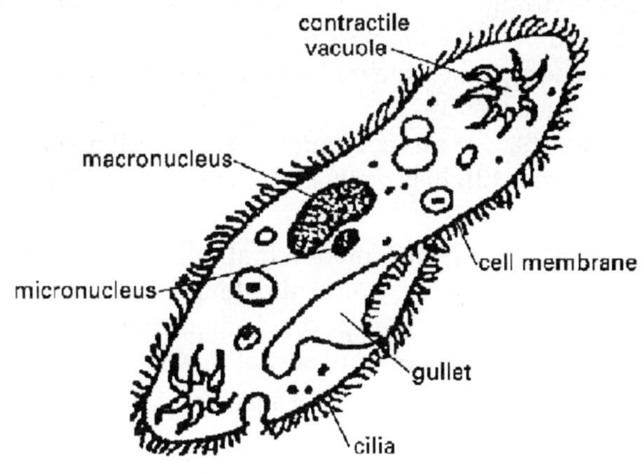

A   B

75. Refer to the illustration above. The process shown in figure B is called _________________.

    Answer: exocytosis   Difficulty: II   Section: 2   Objective: 3

76. Refer to the illustration above. Cells often engulf extracellular particles and fluid, as shown in figure A. This is called _________________.

    Answer: endocytosis   Difficulty: II   Section: 2   Objective: 3

77. Receptor proteins can change the _________________ of the cell membrane.

    Answer: permeability   Difficulty: I   Section: 2   Objective: 4

78. Receptor proteins may act as _________________, catalyzing certain chemical reactions inside the cell.

    Answer: enzymes   Difficulty: I   Section: 2   Objective: 4

79. In the cell membrane, proteins that transmit information into the cell by responding to signal molecules are called _________________.

    Answer: receptors   Difficulty: I   Section: 2   Objective: 4

PROBLEM

Paramecia are unicellular protists. They have a number of characteristics also found in animals, such as the need to ingest food in order to obtain energy (they are heterotrophs), and they are surrounded by a cell membrane but not by a rigid cell wall. They have organelles found in animal cells, including a nucleus, mitochondria, ribosomes, and cilia. In addition, they have star-shaped organelles, called contractile vacuoles, that expel excess water. The illustration below depicts a paramecium.

80. Refer to the illustration above. The data presented in the table below were obtained in an experiment in which paramecia were placed in different salt concentrations and the rate at which the contractile vacuole contracted to pump out excess water was recorded.

Salt	Rate of contractile vacuole contractions/minute
Very high	2
High	8
Medium	15
Low	22
Very low	30

   a. How can you explain the observed relationship between salt concentration and rate of contractile vacuole contraction?

   b. If something happened to a paramecium that caused its contractile vacuole to stop contracting, what would you expect to happen? Would this result occur more quickly if the paramecium was in water with a high salt concentration or in water with a low salt concentration? Why?

Answer:
   a. The contractile vacuole maintains water balance by pumping water out of the cell. When the salt concentration outside of the cell is very high, water will move from inside the cell to outside the cell—little or no pumping action is required. When the salt concentration outside of the cell is low, the tendency is for water to move from outside the cell to inside the cell, necessitating increased pumping action by the vacuole to move excess water out of the cell.
   b. If the contractile vacuole were to stop contracting, the cell would burst because water would collect in excess inside of it and the cell membrane would not be strong enough to resist rupturing. This result would be expected to occur more quickly if the organism were placed in water with a low salt concentration than it would if the organism were placed in water with a high salt concentration. This is because water accumulates inside the paramecium more rapidly when it is placed in a low salt environment.

Difficulty: III   Section: 1   Objective: 2

81. A biologist conducts an experiment designed to determine whether a particular type of molecule is transported into cells by diffusion, facilitated diffusion, or active transport. He collects the following information:
1. The molecule is very small.
2. The molecule is polar.
3. The molecule can accumulate inside a cell, even when its concentration inside the cell initially is higher than it is outside the cell.
4. Cells use up more energy when the molecule is present in the environment around the cells than when it is not present.

The biologist concludes that the molecule moves across cell membranes by facilitated diffusion. Do you agree with his conclusion? Why or why not?

Answer:
Students should disagree. The information that cells can accumulate the molecule against a concentration gradient is compelling evidence that active transport is the mechanism of transport. Active transport also requires energy consumption, which was also found to be a property of transport of this molecule.

Difficulty: III   Section: 2   Objective: 1

ESSAY

82. Why do dissolved particles on one side of a membrane result in the diffusion of water across the membrane?

    Answer:
    Dissolved particles reduce the number of water molecules that can move freely on that side. Water then moves by osmosis from the side where the free water molecule concentration is high to the side where the concentration is lower.

    Difficulty: III   Section: 1   Objective: 2

83. Why is it dangerous for humans to drink sea water?

    Answer:
    The concentration of salt in sea water is higher than the concentration of salt in the fluids that surround the cells in the human body. Drinking sea water increases the concentration of salt in the body's fluids. This causes water to leave the cells by osmosis, and without the proper amount of water, the cells will be harmed or will die.

    Difficulty: III   Section: 1   Objective: 2

84. Distinguish between facilitated diffusion and active transport.

    Answer:
    In facilitated diffusion, carrier proteins assist the diffusion of substances down their concentration gradient. Active transport, on the other hand, is the movement of substances against their concentration gradient.

    Difficulty: II   Section: 2   Objective: 1

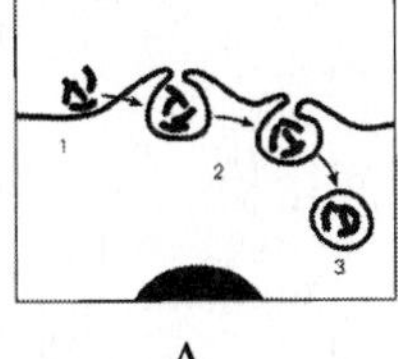

A

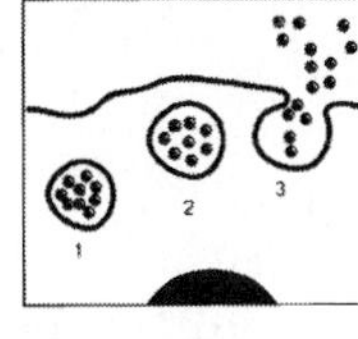

B

85. Refer to the illustration above. Identify and explain the processes taking place in figure A and figure B.

    Answer:
    Endocytosis is the process taking place in figure A. Endocytosis is the process by which cells engulf substances that are too large to enter the cell, by enclosing the substances in vesicles. Exocytosis is the process taking place in figure B. Exocytosis is the process by which cells export substances and discharge waste from vesicles at the cell's surface.

    Difficulty: II   Section: 2   Objective: 3

86. Describe three ways in which the binding of a signal molecule to a receptor protein can change the functioning of a cell.

    Answer:
    1. The receptor can act as an enzyme.
    2. The receptor may cause the formation of a second messenger that will have an effect in another part of the cytoplasm.
    3. The receptor can cause membrane channels to open.

    Difficulty: II   Section: 2   Objective: 4